CAMBRIDGE LIBRARY COLLECTION

Books of enduring scholarly value

Darwin

Two hundred years after his birth and 150 years after the publication of 'On the Origin of Species', Charles Darwin and his theories are still the focus of worldwide attention. This series offers not only works by Darwin, but also the writings of his mentors in Cambridge and elsewhere, and a survey of the impassioned scientific, philosophical and theological debates sparked by his 'dangerous idea'.

Contributions to the Theory of Natural Selection

Alfred Russel Wallace (1823–1913) is regarded as the co-discoverer with Darwin of the theory of evolution. It was an essay which Wallace sent in 1858 to Darwin (to whom he had dedicated his most famous book, The Malay Archipelago) which impelled Darwin to publish an article on his own long-pondered theory simultaneously with that of Wallace. As a travelling naturalist and collector in the Far East and South America, Wallace already inclined towards the Lamarckian theory of transmutation of species, and his own researches convinced him of the reality of evolution. On the publication of On the Origin of Species, Wallace became one of its most prominent advocates. This second, corrected, edition (1871) of a series of essays published in book form in 1870, shows the development of his thinking about evolution, and emphasises his admiration for, and support of, Darwin's work. As he says in the Preface: 'I have felt all my life, and I still feel, the most sincere satisfaction that Mr Darwin had been at work long before me, and that it was not left for me to attempt to write The Origin of Species. I have long since measured my own strength, and know well that it would be quite unequal to the task. Far abler men than myself may confess, that they have not that untiring patience in accumulating, and that wonderful skill in using, large masses of facts of the most varied kind, that wide and accurate physiological knowledge, that acuteness in devising and skill in carrying out experiments, and that admirable style of composition, at once clear, persuasive and judicial – qualities which, in their harmonious combination mark out Mr Darwin as the man, perhaps of all men now living, best fitted for the great work he has undertaken and accomplished.'

T0273858

Cambridge University Press has long been a pioneer in the reissuing of out-of-print titles from its own backlist, producing digital reprints of books that are still sought after by scholars and students but could not be reprinted economically using traditional technology. The Cambridge Library Collection extends this activity to a wider range of books which are still of importance to researchers and professionals, either for the source material they contain, or as landmarks in the history of their academic discipline.

Drawing from the world-renowned collections in the Cambridge University Library, and guided by the advice of experts in each subject area, Cambridge University Press is using state-of-the-art scanning machines in its own Printing House to capture the content of each book selected for inclusion. The files are processed to give a consistently clear, crisp image, and the books finished to the high quality standard for which the Press is recognised around the world. The latest print-on-demand technology ensures that the books will remain available indefinitely, and that orders for single or multiple copies can quickly be supplied.

The Cambridge Library Collection will bring back to life books of enduring scholarly value (including out-of-copyright works originally issued by other publishers) across a wide range of disciplines in the humanities and social sciences and in science and technology.

Contributions to the Theory of Natural Selection

A Series of Essays

ALFRED RUSSEL WALLACE

CAMBRIDGE UNIVERSITY PRESS

Cambridge, New York, Melbourne, Madrid, Cape Town, Singapore,
São Paolo, Delhi, Dubai, Tokyo

Published in the United States of America by Cambridge University Press, New York

www.cambridge.org
Information on this title: www.cambridge.org/9781108001540

© in this compilation Cambridge University Press 2009

This edition first published 1871
This digitally printed version 2009

ISBN 978-1-108-00154-0 Paperback

CONTRIBUTIONS TO

THE THEORY OF

NATURAL SELECTION.

𝕬 𝕾𝖊𝖗𝖎𝖊𝖘 𝖔𝖋 𝕰𝖘𝖘𝖆𝖞𝖘.

BY

ALFRED RUSSEL WALLACE,

AUTHOR OF

"THE MALAY ARCHIPELAGO," ETC., ETC.

SECOND EDITION, WITH CORRECTIONS AND ADDITIONS.

London:

MACMILLAN AND CO.,

1871.

PREFACE.

THE present volume consists of essays which I have contributed to various periodicals, or read before scientific societies during the last fifteen years, with others now printed for the first time. The two first of the series are printed without alteration, because, having gained me the reputation of being an independent originator of the theory of "natural selection," they may be considered to have some historical value. I have added to them one or two very short explanatory notes, and have given headings to subjects, to make them uniform with the rest of the book. The other essays have been carefully corrected, often considerably enlarged, and in some cases almost rewritten, so as to express more fully and more clearly the views which I hold at the present time; and as most of them originally appeared in publications which have a very limited circulation, I believe that the larger portion of this volume will be new to many of my friends and to most of my readers.

I now wish to say a few words on the reasons which have led me to publish this work. The second essay, especially when taken in connection with the first, contains an outline sketch of the theory of the origin of species (by means of what was afterwards termed by Mr. Darwin—"natural selection,") as conceived by me before I had the least notion of the scope and nature of Mr. Darwin's labours. They were published in a way not likely to attract the attention of any but working naturalists, and I feel sure that many who have heard of them, have never had the opportunity of ascertaining how much or how little they really contain. It therefore happens, that, while some writers give me more credit than I deserve, others may very naturally class me with Dr. Wells and Mr. Patrick Matthew, who, as Mr. Darwin has shown in the historical sketch given in the 4th and 5th Editions of the "Origin of Species," certainly propounded the fundamental principle of "natural selection" before himself, but who made no further use of that principle, and failed to see its wide and immensely important applications.

The present work will, I venture to think, prove, that I both saw at the time the value and scope of the law which I had discovered, and have since been able to apply it to some purpose in a few original lines of investigation. But here my claims cease. I have felt all my life, and I still feel, the most sincere satisfaction that Mr. Darwin had been at work long before me, and that it was not left for me to attempt to write "The Origin of Species." I have long since measured my own strength, and know well that it would be quite unequal to that task. Far abler men than myself may confess, that they have not that untiring patience in accumulating, and that wonderful skill in using, large masses of facts of the most varied kind,—that wide and accurate physiological knowledge,—that acuteness in devising and skill in carrying out experiments,—and that admirable style of composition, at once clear, persuasive and judicial,—qualities, which in their harmonious combination mark out Mr. Darwin as the man, perhaps of all men now living, best fitted for the great work he has undertaken and accomplished.

My own more limited powers have, it is true, enabled me now and then to seize on some conspicuous group of unappropriated facts, and to search out some generalization which might bring them under the reign of known law; but they are not suited to that more scientific and more laborious process of elaborate induction, which in Mr. Darwin's hands has led to such brilliant results.

Another reason which has led me to publish this volume at the present time is, that there are some important points on which I differ from Mr. Darwin, and I wish to put my opinions on record in an easily accessible form, before the publication of his new work, (already announced,) in which I believe most of these disputed questions will be fully discussed.

I will now give the date and mode of publication of each of the essays in this volume, as well as the amount of alteration they have undergone.

I.—ON THE LAW WHICH HAS REGULATED THE INTRODUCTION OF NEW SPECIES.

First published in the " Annals and Magazine of

Natural History," September, 1855. Reprinted without alteration of the text.

II.—On the Tendency of Varieties to Depart indefinitely from the Original Type.

First published in the " Journal of the Proceedings of the Linnæan Society," August, 1858. Reprinted without alteration of the text, except one or two grammatical emendations.

III.—Mimicry and other Protective Resemblances among Animals.

First published in the " Westminster Review," July, 1867. Reprinted with a few corrections and some important additions, among which I may especially mention Mr. Jenner Weir's observations and experiments on the colours of the caterpillars eaten or rejected by birds.

IV.—The Malayan Papilionidæ, or Swallow-Tailed Butterflies, as Illustrative of the Theory of Natural Selection.

First published in the " Transactions of the Linnæan Society," Vol. XXV. (read March, 1864), under the title, " On the Phenomena of Variation and Geographical Distribution, as illustrated by the Papilionidæ of the Malayan Region."

The introductory part of this essay is now reprinted, omitting tables, references to plates, &c., with some additions, and several corrections. Owing to the publication of Dr. Felder's "Voyage of the Novara" (Lepidoptera) in the interval between the reading of my paper and its publication, several of my new species must have their names changed for those given to them by Dr. Felder, and this will explain the want of agreement in some cases between the names used in this volume and those of the original paper.

V.—On Instinct in Man and Animals.

Not previously published.

VI.—The Philosophy of Birds' Nests.

First published in the "Intellectual Observer," July, 1867. Reprinted with considerable emendations and additions.

VII.—A Theory of Birds' Nests;

Showing the relation of certain differences of Colour in Birds to their mode of Nidification.

First published in the "Journal of Travel and Natural History" (No. 2), 1868. Now reprinted with considerable emendations and additions, by which I have endeavoured more clearly to express, and more fully to illustrate, my meaning in those parts which have been misunderstood by my critics.

VIII.—Creation by Law.

First published in the "Quarterly Journal of Science," October, 1867. Now reprinted with a few alterations and additions.

IX.—The Development of Human Races under the Law of Natural Selection.

First published in the "Anthropological Review," May, 1864. Now reprinted with a few important alterations and additions. I had intended to have considerably extended this essay, but on attempting it I found that I should probably weaken the effect without adding much to the argument. I have therefore preferred to leave it as it was first written, with the exception of a few ill-considered passages which never fully expressed my meaning. As it now stands, I believe it contains the enunciation of an important truth.

X.—The Limits of Natural Selection as applied to Man.

This is the further development of a few sentences at the end of an article on "Geological Time and the Origin of Species," which appeared in the "Quarterly Review," for April, 1869. I have here ventured to touch on a class of problems which are usually considered to be beyond the boundaries of science, but which, I believe, will one day be brought within her domain.

For the convenience of those who are acquainted with any of my essays in their original form, I subjoin references to the more important additions and alterations now made to them.

*ADDITIONS AND CORRECTIONS TO THE ESSAYS AS
ORIGINALLY PUBLISHED.*

Essays I. and II. are unaltered, but short notes are
added at pp. 19, 24, 29, and 40.

III.—*Mimicry, and other Protective Resemblances
among Animals.*

PAGE

IV.—*The Malayan Papilionidæ or Swallow-tailed
Butterflies, as illustrative of the Theory of Natural
Selection.*

London, March, 1870.

PREFACE TO THE SECOND EDITION.

THE flattering reception of my Essays by the public and the press having led to a second edition being called for within a year of its first publication, I have taken the opportunity to make a few necessary corrections. I have also added a few passages to the 6th and 7th Essays, and have given two notes, explanatory of some portions of the last chapter which appear to have been not always understood. These additions are as follows :—

To avoid altering the paging the additional pages now given have been lettered.

1st Ed.	2nd Ed.	
221	221	Additional facts as to birds acquiring the song of other species.
223	223A 223B	Mr. Spruce's remarks on young birds pairing with old.
228	228A 228B	Pouchet's observations on a change in the nests of swallows.
229	—	Passage omitted about nest of Golden Crested Warbler, which had been inserted on Rennie's authority, but has not been confirmed by any later observers.
261	261	Daines Barrington, on importance of protection to the female bird.
	372	Note A.
	372B	Note B.

CONTENTS.

VIII.—*Creation by Law.*

IX.—*The Development of Human Races under the Law of Natural Selection.*

X.—*The Limits of Natural Selection as applied to Man.*

I.

ON THE LAW WHICH HAS REGULATED THE INTRODUCTION OF NEW SPECIES.*

Geographical Distribution dependent on Geologic Changes.

EVERY naturalist who has directed his attention to the subject of the geographical distribution of animals and plants, must have been interested in the singular facts which it presents. Many of these facts are quite different from what would have been anticipated, and have hitherto been considered as highly curious, but quite inexplicable. None of the explanations attempted from the time of Linnæus are now considered at all satisfactory; none of them have given a cause sufficient to account for the facts known at the time, or comprehensive enough to include all the new facts which have since been, and are daily being added. Of late years, however, a great light has been thrown upon the subject by geological investigations, which have shown that the present state of the earth and of the organisms now

---

* Written at Sarawak in February, 1855, and published in the " Annals and Magazine of Natural History," September, 1855.

B

inhabiting it, is but the last stage of a long and uninterrupted series of changes which it has undergone, and consequently, that to endeavour to explain and account for its present condition without any reference to those changes (as has frequently been done) must lead to very imperfect and erroneous conclusions.

The facts proved by geology are briefly these :— That during an immense, but unknown period, the surface of the earth has undergone successive changes; land has sunk beneath the ocean, while fresh land has risen up from it; mountain chains have been elevated ; islands have been formed into continents, and continents submerged till they have become islands ; and these changes have taken place, not once merely, but perhaps hundreds, perhaps thousands of times :—That all these operations have been more or less continuous, but unequal in their progress, and during the whole series the organic life of the earth has undergone a corresponding alteration.   This alteration also has been gradual, but complete ; after a certain interval not a single species existing which had lived at the commencement of the period.   This complete renewal of the forms of life also appears to have occurred several times :—That from the last of the geological epochs to the present or historical epoch, the change of organic life has been gradual : the first appearance of animals now existing can in many cases be traced, their numbers gradually increasing in the more recent formations, while other species continually die out and disappear, so that the present condition of the organic world is clearly derived by a natural process of gradual extinction and creation of species from that of the latest geological periods. We may therefore safely infer a like gradation and natural sequence from one geological epoch to another.

Now, taking this as a fair statement of the results of geological inquiry, we see that the present geographical distribution of life upon the earth must be the result of all the previous changes, both of the surface of the earth itself and of its inhabitants. Many causes, no doubt, have operated of which we must ever remain in ignorance, and we may, therefore, expect to find many details very difficult of explanation, and in attempting to give one, must allow ourselves to call into our service geological changes which it is highly probable may have occurred, though we have no direct evidence of their individual operation.

The great increase of our knowledge within the last twenty years, both of the present and past history of the organic world, has accumulated a body of facts which should afford a sufficient foundation for a comprehensive law embracing and explaining them all, and giving a direction to new researches. It is about ten years since the idea of such a law suggested itself to the writer of this essay, and he has since taken every opportunity of testing it by all the newly-ascertained facts with which he has become

acquainted, or has been able to observe himself.
These have all served to convince him of the correctness of his hypothesis. Fully to enter into such a
subject would occupy much space, and it is only in
consequence of some views having been lately promulgated, he believes, in a wrong direction, that he
now ventures to present his ideas to the public, with
only such obvious illustrations of the arguments and
results as occur to him in a place far removed from
all means of reference and exact information.

*A Law deduced from well-known Geographical and
Geological Facts.*

The following propositions in Organic Geography
and Geology give the main facts on which the
hypothesis is founded.

Geography.

1. Large groups, such as classes and orders, are
generally spread over the whole earth, while smaller
ones, such as families and genera, are frequently
confined to one portion, often to a very limited district.

2. In widely distributed families the genera are
often limited in range ; in widely distributed genera,
well marked groups of species are peculiar to each
geographical district.

3. When a group is confined to one district, and
is rich in species, it is almost invariably the case
that the most closely allied species are found in the
same locality or in closely adjoining localities, and

that therefore the natural sequence of the species by affinity is also geographical.

4. In countries of a similar climate, but separated by a wide sea or lofty mountains, the families, genera and species of the one are often represented by closely allied families, genera and species peculiar to the other.

Geology.

5. The distribution of the organic world in time is very similar to its present distribution in space.

6. Most of the larger and some small groups extend through several geological periods.

7. In each period, however, there are peculiar groups, found nowhere else, and extending through one or several formations.

8. Species of one genus, or genera of one family occurring in the same geological time are more closely allied than those separated in time.

9. As generally in geography no species or genus occurs in two very distant localities without being also found in intermediate places, so in geology the life of a species or genus has not been interrupted. In other words, no group or species has come into existence twice.

10. The following law may be deduced from these facts :—*Every species has come into existence coincident both in space and time with a pre-existing closely allied species.*

This law agrees with, explains and illustrates all the facts connected with the following branches of the subject :—1st. The system of natural affinities. 2nd. The distribution of animals and plants in space. 3rd. The same in time, including all the phænomena of representative groups, and those which Professor Forbes supposed to manifest polarity. 4th. The phænomena of rudimentary organs. We will briefly endeavour to show its bearing upon each of these.

The Form of a true system of Classification determined by this Law.

If the law above enunciated be true, it follows that the natural series of affinities will also represent the order in which the several species came into existence, each one having had for its immediate antitype a closely allied species existing at the time of its origin. It is evidently possible that two or three distinct species may have had a common antitype, and that each of these may again have become the antitypes from which other closely allied species were created. The effect of this would be, that so long as each species has had but one new species formed on its model, the line of affinities will be simple, and may be represented by placing the several species in direct succession in a straight line. But if two or more species have been independently formed on the plan of a common antitype, then the series of affinities will be compound, and can only be represented by a forked or many branched line. Now, all attempts at a Natural classification and arrangement of organic beings show, that both these plans have obtained in creation. Sometimes the series of affinities can be well represented for a space by a direct progression from species to species or from group to group, but it is generally found impossible so to continue. There constantly occur two or more modifications of an organ or modifications of two distinct organs, leading us on to two distinct series of species, which at length differ so much from each other as to form distinct genera or families. These are the parallel series or representative groups of naturalists, and they often occur in different countries, or are found fossil in different formations. They are said to have an analogy to each other when they are so far removed from their common antitype as to differ in many important points of structure, while they still preserve a family resemblance. We thus see how difficult it is to determine in every case whether a given relation is an analogy or an affinity, for it is evident that as we go back along the parallel or divergent series, towards the common antitype, the analogy which existed between the two groups becomes an affinity. We are also made aware of the difficulty of arriving at a true classification, even in a small and perfect group ;—in the actual state of nature it is almost impossible, the species being so numerous and the modifications of form and structure so varied, arising probably from the immense number of species which have served as antitypes for the existing species, and thus produced a complicated branching of

the lines of affinity, as intricate as the twigs of a
gnarled oak or the vascular system of the human
body. Again, if we consider that we have only fragments of this vast system, the stem and main branches
being represented by extinct species of which we have
no knowledge, while a vast mass of limbs and boughs
and minute twigs and scattered leaves is what we
have to place in order, and determine the true position each originally occupied with regard to the others,
the whole difficulty of the true Natural System of
classification becomes apparent to us.

We shall thus find ourselves obliged to reject
all those systems of classification which arrange
species or groups in circles, as well as those
which fix a definite number for the divisions of
each group. The latter class have been very generally rejected by naturalists, as contrary to nature,
notwithstanding the ability with which they have
been advocated ; but the circular system of affinities seems to have obtained a deeper hold, many
eminent naturalists having to some extent adopted
it. We have, however, never been able to find a
case in which the circle has been closed by a
direct and close affinity. In most cases a palpable
analogy has been substituted, in others the affinity
is very obscure or altogether doubtful. The complicated branching of the lines of affinities in
extensive groups must also afford great facilities
for giving a show of probability to any such
purely artificial arrangements. Their death-blow

was given by the admirable paper of the lamented Mr. Strickland, published in the "Annals of Natural History," in which he so clearly showed the true synthetical method of discovering the Natural System.

Geographical Distribution of Organisms.

If we now consider the geographical distribution of animals and plants upon the earth, we shall find all the facts beautifully in accordance with, and readily explained by, the present hypothesis. A country having species, genera, and whole families peculiar to it, will be the necessary result of its having been isolated for a long period, sufficient for many series of species to have been created on the type of pre-existing ones, which, as well as many of the earlier-formed species, have become extinct, and thus made the groups appear isolated. If in any case the antitype had an extensive range, two or more groups of species might have been formed, each varying from it in a different manner, and thus producing several representative or analogous groups. The Sylviadæ of Europe and the Sylvicolidæ of North America, the Heliconidæ of South America and the Euplœas of the East, the group of Trogons inhabiting Asia, and that peculiar to South America, are examples that may be accounted for in this manner.

Such phænomena as are exhibited by the Gala-

pagos Islands, which contain little groups of plants
and animals peculiar to themselves, but most nearly
allied to those of South America, have not hitherto received any, even a conjectural explanation.
The Galapagos are a volcanic group of high antiquity, and have probably never been more closely
connected with the continent than they are at
present. They must have been first peopled, like
other newly-formed islands, by the action of winds
and currents, and at a period sufficiently remote
to have had the original species die out, and the
modified prototypes only remain. In the same way
we can account for the separate islands having each
their peculiar species, either on the supposition that
the same original emigration peopled the whole of
the islands with the same species from which differently modified prototypes were created, or that the
islands were successively peopled from each other,
but that new species have been created in each on
the plan of the pre-existing ones. St. Helena is a
similar case of a very ancient island having obtained
an entirely peculiar, though limited, flora. On the
other hand, no example is known of an island which
can be proved geologically to be of very recent
origin (late in the Tertiary, for instance), and yet
possesses generic or family groups, or even many
species peculiar to itself.

When a range of mountains has attained a great
elevation, and has so remained during a long geological period, the species of the two sides at and

near their bases will be often very different, representative species of some genera occurring, and even whole genera being peculiar to one side only, as is remarkably seen in the case of the Andes and Rocky Mountains. A similar phænomenon occurs when an island has been separated from a continent at a very early period. The shallow sea between the Peninsula of Malacca, Java, Sumatra and Borneo was probably a continent or large island at an early epoch, and may have become submerged as the volcanic ranges of Java and Sumatra were elevated. The organic results we see in the very considerable number of species of animals common to some or all of these countries, while at the same time a number of closely allied representative species exist peculiar to each, showing that a considerable period has elapsed since their separation. The facts of geographical distribution and of geology may thus mutually explain each other in doubtful cases, should the principles here advocated be clearly established.

In all those cases in which an island has been separated from a continent, or raised by volcanic or coralline action from the sea, or in which a mountain-chain has been elevated in a recent geological epoch, the phænomena of peculiar groups or even of single representative species will not exist. Our own island is an example of this, its separation from the continent being geologically very recent, and we have consequently scarcely a species which is peculiar to it; while the Alpine range, one of the most recent mountain elevations, separates faunas and floras which scarcely differ more than may be due to climate and latitude alone.

The series of facts alluded to in Proposition (3), of closely allied species in rich groups being found geographically near each other, is most striking and important. Mr. Lovell Reeve has well exemplified it in his able and interesting paper on the Distribution of the Bulimi. It is also seen in the Humming-birds and Toucans, little groups of two or three closely allied species being often found in the same or closely adjoining districts, as we have had the good fortune of personally verifying. Fishes give evidence of a similar kind : each great river has its peculiar genera, and in more extensive genera its groups of closely allied species. But it is the same throughout Nature; every class and order of animals will contribute similar facts. Hitherto no attempt has been made to explain these singular phænomena, or to show how they have arisen. Why are the genera of Palms and of Orchids in almost every case confined to one hemisphere ? Why are the closely allied species of brown-backed Trogons all found in the East, and the green-backed in the West? Why are the Macaws and the Cockatoos similarly restricted ? Insects furnish a countless number of analogous examples ;—the Goliathi of Africa, the Ornithopteræ of the Indian Islands, the Heliconidæ of South America, the Danaidæ of the East, and in all, the most closely allied species found in geographical proximity. The question forces itself upon every thinking mind,—why are these things so ? They could not be as they are had no law regulated their creation and dispersion. The law here enunciated not merely explains, but necessitates the facts we see to exist, while the vast and long-continued geological changes of the earth readily account for the exceptions and apparent discrepancies that here and there occur. The writer's object in putting forward his views in the present imperfect manner is to submit them to the test of other minds, and to be made aware of all the facts supposed to be inconsistent with them. As his hypothesis is one which claims acceptance solely as explaining and connecting facts which exist in nature, he expects facts alone to be brought to disprove it, not à priori arguments against its probability.

Geological Distribution of the Forms of Life.

The phænomena of geological distribution are exactly analogous to those of geography. Closely allied species are found associated in the same beds, and the change from species to species appears to have been as gradual in time as in space. Geology, however, furnishes us with positive proof of the extinction and production of species, though it does not inform us how either has taken place. The extinction of species, however, offers but little difficulty, and the *modus operandi* has been well illustrated by Sir

C. Lyell in his admirable "Principles." Geological changes, however gradual, must occasionally have modified external conditions to such an extent as to have rendered the existence of certain species impossible. The extinction would in most cases be effected by a gradual dying-out, but in some instances there might have been a sudden destruction of a species of limited range. To discover how the extinct species have from time to time been replaced by new ones down to the very latest geological period, is the most difficult, and at the same time the most interesting problem in the natural history of the earth. The present inquiry, which seeks to eliminate from known facts a law which has determined, to a certain degree, what species could and did appear at a given epoch, may, it is hoped, be considered as one step in the right direction towards a complete solution of it.

High Organization of very ancient Animals consistent with this Law.

Much discussion has of late years taken place on the question, whether the succession of life upon the globe has been from a lower to a higher degree of organization. The admitted facts seem to show that there has been a general, but not a detailed progression. Mollusca and Radiata existed before Vertebrata, and the progression from Fishes to Reptiles and Mammalia, and also from the lower mammals to the higher, is indisputable. On the other hand,

it is said that the Mollusca and Radiata of the very
earliest periods were more highly organized than
the great mass of those now existing, and that the
very first fishes that have been discovered are by no
means the lowest organised of the class. Now it is
believed the present hypothesis will harmonize with
all these facts, and in a great measure serve to
explain them ; for though it may appear to some
readers essentially a theory of progression, it is in
reality only one of gradual change. It is, however,
by no means difficult to show that a real progression
in the scale of organization is perfectly consistent
with all the appearances, and even with apparent
retrogression, should such occur.

Returning to the analogy of a branching tree, as
the best mode of representing the natural arrangement of species and their successive creation, let us
suppose that at an early geological epoch any group
(say a class of the Mollusca) has attained to a great
richness of species and a high organization. Now
let this great branch of allied species, by geological mutations, be completely or partially destroyed.
Subsequently a new branch springs from the same
trunk, that is to say, new species are successively
created, having for their antitypes the same lower
organized species which had served as the antitypes
for the former group, but which have survived the
modified conditions which destroyed it. This new
group being subject to these altered conditions, has
modifications of structure and organization given

to it, and becomes the representative group of the former one in another geological formation. It may, however, happen, that though later in time, the new series of species may never attain to so high a degree of organization as those preceding it, but in its turn become extinct, and give place to yet another modification from the same root, which may be of higher or lower organization, more or less numerous in species, and more or less varied in form and structure than either of those which preceded it. Again, each of these groups may not have become totally extinct, but may have left a few species, the modified prototypes of which have existed in each succeeding period, a faint memorial of their former grandeur and luxuriance. Thus every case of apparent retrogression may be in reality a progress, though an interrupted one : when some monarch of the forest loses a limb, it may be replaced by a feeble and sickly substitute. The foregoing remarks appear to apply to the case of the Mollusca, which, at a very early period, had reached a high organization and a great development of forms and species in the testaceous Cephalopoda. In each succeeding age modified species and genera replaced the former ones which had become extinct, and as we approach the present æra, but few and small representatives of the group remain, while the Gasteropods and Bivalves have acquired an immense preponderance. In the long series of changes the earth has undergone, the process of peopling it with organic beings has been continually going on, and whenever any of the higher groups have become nearly or quite extinct, the lower forms which have better resisted the modified physical conditions have served as the antitypes on which to found the new races. In this manner alone, it is believed, can the representative groups at successive periods, and the risings and fallings in the scale of organization, be in every case explained.

Objections to Forbes' Theory of Polarity.

The hypothesis of polarity, recently put forward by Professor Edward Forbes to account for the abundance of generic forms at a very early period and at present, while in the intermediate epochs there is a gradual diminution and impoverishment, till the minimum occurred at the confines of the Palæozoic and Secondary epochs, appears to us quite unnecessary, as the facts may be readily accounted for on the principles already laid down. Between the Palæozoic and Neozoic periods of Professor Forbes, there is scarcely a species in common, and the greater part of the genera and families also disappear to be replaced by new ones. It is almost universally admitted that such a change in the organic world must have occupied a vast period of time. Of this interval we have no record; probably because the whole area of the early formations now exposed to our researches was elevated at the end of the Palæozoic period, and remained so through the interval required for the organic changes which

c

resulted in the fauna and flora of the Secondary period. The records of this interval are buried beneath the ocean which covers three-fourths of the globe. Now it appears highly probable that a long period of quiescence or stability in the physical conditions of a district would be most favourable to the existence of organic life in the greatest abundance, both as regards individuals and also as to variety of species and generic group, just as we now find that the places best adapted to the rapid growth and increase of individuals also contain the greatest profusion of species and the greatest variety of forms, —the tropics in comparison with the temperate and arctic regions. On the other hand, it seems no less probable that a change in the physical conditions of a district, even small in amount if rapid, or even gradual if to a great amount, would be highly unfavourable to the existence of individuals, might cause the extinction of many species, and would probably be equally unfavourable to the creation of new ones. In this too we may find an analogy with the present state of our earth, for it has been shown to be the violent extremes and rapid changes of physical conditions, rather than the actual mean state in the temperate and frigid zones, which renders them less prolific than the tropical regions, as exemplified by the great distance beyond the tropics to which tropical forms penetrate when the climate is equable, and also by the richness in species and forms of tropical mountain regions which principally differ from the temperate zone in the uniformity of their climate. However this may be, it seems a fair assumption that during a period of geological repose the new species which we know to have been created would have appeared, that the creations would then exceed in number the extinctions, and therefore the number of species would increase. In a period of geological activity, on the other hand, it seems probable that the extinctions might exceed the creations, and the number of species consequently diminish. That such effects did take place in connexion with the causes to which we have imputed them, is shown in the case of the Coal formation, the faults and contortions of which show a period of great activity and violent convulsions, and it is in the formation immediately succeeding this that the poverty of forms of life is most apparent. We have then only to suppose a long period of somewhat similar action during the vast unknown interval at the termination of the Palæozoic period, and then a decreasing violence or rapidity through the Secondary period, to allow for the gradual repopulation of the earth with varied forms, and the whole of the facts are explained.* We thus have a clue to the increase of the forms of life during certain periods, and their decrease during others, without recourse

* Professor Ramsay has since shown that a glacial epoch probably occurred at the time of the Permian formation, which will more satisfactorily account for the comparative poverty of species.

to any causes but those we know to have existed, and to effects fairly deducible from them. The precise manner in which the geological changes of the early formations were effected is so extremely obscure, that when we can explain important facts by a retardation at one time and an acceleration at another of a process which we know from its nature and from observation to have been unequal,—a cause so simple may surely be preferred to one so obscure and hypothetical as polarity.

I would also venture to suggest some reasons against the very nature of the theory of Professor Forbes. Our knowledge of the organic world during any geological epoch is necessarily very imperfect. Looking at the vast numbers of species and groups that have been discovered by geologists, this may be doubted; but we should compare their numbers not merely with those that now exist upon the earth, but with a far larger amount. We have no reason for believing that the number of species on the earth at any former period was much less than at present; at all events the aquatic portion, with which geologists have most acquaintance, was probably often as great or greater. Now we know that there have been many complete changes of species; new sets of organisms have many times been introduced in place of old ones which have become extinct, so that the total amount which have existed on the earth from the earliest geological period must have borne about the same proportion to those now living, as the whole human race who have lived and died upon the earth, to the population at the present time. Again, at each epoch, the whole earth was no doubt, as now, more or less the theatre of life, and as the successive generations of each species died, their exuviæ and preservable parts would be deposited over every portion of the then existing seas and oceans, which we have reason for supposing to have been more, rather than less, extensive than at present. In order then to understand our possible knowledge of the early world and its inhabitants, we must compare, not the area of the whole field of our geological researches with the earth's surface, but the area of the examined portion of each formation separately with the whole earth. For example, during the Silurian period all the earth was Silurian, and animals were living and dying, and depositing their remains more or less over the whole area of the globe, and they were probably (the species at least) nearly as varied in different latitudes and longitudes as at present. What proportion do the Silurian districts bear to the whole surface of the globe, land and sea (for far more extensive Silurian districts probably exist beneath the ocean than above it), and what portion of the known Silurian districts has been actually examined for fossils? Would the area of rock actually laid open to the eye be the thousandth or the ten-thousandth part of the earth's surface? Ask the same question with regard to the Oolite or the Chalk, or even to particular beds of these when they differ considerably in their fossils, and you may then get some notion of how small a portion of the whole we know.

But yet more important is the probability, nay almost the certainty, that whole formations containing the records of vast geological periods are entirely buried beneath the ocean, and for ever beyond our reach. Most of the gaps in the geological series may thus be filled up, and vast numbers of unknown and unimaginable animals, which might help to elucidate the affinities of the numerous isolated groups which are a perpetual puzzle to the zoologist, may there be buried, till future revolutions may raise them in their turn above the waters, to afford materials for the study of whatever race of intelligent beings may then have succeeded us. These considerations must lead us to the conclusion, that our knowledge of the whole series of the former inhabitants of the earth is necessarily most imperfect and fragmentary,—as much so as our knowledge of the present organic world would be, were we forced to make our collections and observations only in spots equally limited in area and in number with those actually laid open for the collection of fossils. Now, the hypothesis of Professor Forbes is essentially one that assumes to a great extent the completeness of our knowledge of the whole series of organic beings which have existed on the earth. This appears to be a fatal objection to it, independently of all other considerations. It may be said that the same objections exist against every theory on such a subject, but this is not necessarily the case. The hypothesis put forward in this paper depends in no degree upon the completeness of our knowledge of the former condition of the organic world, but takes what facts we have as fragments of a vast whole, and deduces from them something of the nature and proportions of that whole which we can never know in detail. It is founded upon isolated groups of facts, recognizes their isolation, and endeavours to deduce from them the nature of the intervening portions.

Rudimentary Organs

Another important series of facts, quite in accordance with, and even necessary deductions from, the law now developed, are those of rudimentary organs. That these really do exist, and in most cases have no special function in the animal œconomy, is admitted by the first authorities in comparative anatomy. The minute limbs hidden beneath the skin in many of the snake-like lizards, the anal hooks of the boa constrictor, the complete series of jointed finger-bones in the paddle of the Manatus and whale, are a few of the most familiar instances. In botany a similar class of facts has been long recognised. Abortive stamens, rudimentary floral envelopes and undeveloped carpels, are of the most frequent occurrence. To every thoughtful naturalist the question must arise, What are these for? What have they to do with the great laws of creation?

Do they not teach us something of the system or Nature? If each species has been created independently, and without any necessary relations with pre-existing species, what do these rudiments, these apparent imperfections mean? There must be a cause for them; they must be the necessary results of some great natural law. Now, if, as it has been endeavoured to be shown, the great law which has regulated the peopling of the earth with animal and vegetable life is, that every change shall be gradual; that no new creature shall be formed widely differing from anything before existing; that in this, as in everything else in Nature, there shall be gradation and harmony,—then these rudimentary organs are necessary, and are an essential part of the system of Nature. Ere the higher Vertebrata were formed, for instance, many steps were required, and many organs had to undergo modifications from the rudimental condition in which only they had as yet existed. We still see remaining an antitypal sketch of a wing adapted for flight in the scaly flapper of the penguin, and limbs first concealed beneath the skin, and then weakly protruding from it, were the necessary gradations before others should be formed fully adapted for locomotion.* Many more of these modificat tions should we behold, and more complete series

* The theory of Natural Selection has now taught us that these are not the steps by which limbs have been formed; and that most rudimentary organs have been produced by abortion, owing to disuse, as explained by Mr. Darwin.

of them, had we a view of all the forms which have ceased to live. The great gaps that exist between fishes, reptiles, birds, and mammals would then, no doubt, be softened down by intermediate groups, and the whole organic world would be seen to be an unbroken and harmonious system

Conclusion.

It has now been shown, though most briefly and imperfectly, how the law that *"Every species has come into existence coincident both in time and space with a pre-existing closely allied species,"* connects together and renders intelligible a vast number of independent and hitherto unexplained facts. The natural system of arrangement of organic beings, their geographical distribution, their geological sequence, the phænomena of representative and substituted groups in all their modifications, and the most singular peculiarities of anatomical structure, are all explained and illustrated by it, in perfect accordance with the vast mass of facts which the researches of modern naturalists have brought together, and, it is believed, not materially opposed to any of them. It also claims a superiority over previous hypotheses, on the ground that it not merely explains, but necessitates what exists. Granted the law, and many of the most important facts in Nature could not have been otherwise, but are almost as necessary deductions from it, as are the elliptic orbits of the planets from the law of gravitation.

II.

ON THE TENDENCY OF VARIETIES TO DEPART INDEFINITELY FROM THE ORIGINAL TYPE.*

Instability of Varieties supposed to prove the permanent distinctness of Species.

ONE of the strongest arguments which have been adduced to prove the original and permanent distinctness of species is, that *varieties* produced in a state of domesticity are more or less unstable, and often have a tendency, if left to themselves, to return to the normal form of the parent species; and this instability is considered to be a distinctive peculiarity of all varieties, even of those occurring among wild animals in a state of nature, and to constitute a provision for preserving unchanged the originally created distinct species.

In the absence or scarcity of facts and observations as to *varieties* occurring among wild animals, this argument has had great weight with naturalists, and has led to a very general and somewhat

* Written at Ternate, February, 1858; and published in the Journal of the Proceedings of the Linnæan Society for August, 1858.

prejudiced belief in the stability of species. Equally general, however, is the belief in what are called " permanent or true varieties,"—races of animals which continually propagate their like, but which differ so slightly (although constantly) from some other race, that the one is considered to be a *variety* of the other. Which is the *variety* and which the original *species*, there is generally no means of determining, except in those rare cases in which the one race has been known to produce an offspring unlike itself and resembling the other. This, however, would seem quite incompatible with the " permanent invariability of species," but the difficulty is overcome by assuming that such varieties have strict limits, and can never again vary further from the original type, although they may return to it, which, from the analogy of the domesticated animals, is considered to be highly probable, if not certainly proved.

It will be observed that this argument rests entirely on the assumption, that *varieties* occurring in a state of nature are in all respects analogous to or even identical with those of domestic animals, and are governed by the same laws as regards their permanence or further variation. But it is the object of the present paper to show that this assumption is altogether false, that there is a general principle in nature which will cause many *varieties* to survive the parent species, and to give rise to successive variations departing further and further from the original type, and which also produces, in **domesti-cated** animals, the tendency of varieties to **return** to the parent form.

The Struggle for Existence.

The life of wild animals is a struggle for exist-ence. The full exertion of all their faculties and all their energies is required to preserve their own existence and provide for that of their infant off-spring. The possibility of procuring food during the least favourable seasons, and of escaping the attacks of their most dangerous enemies, are the primary conditions which determine the existence both of individuals and of entire species. These conditions will also determine the population of a species; and by a careful consideration of all the circumstances we may be enabled to comprehend, and in some degree to explain, what at first sight appears so inex-plicable—the excessive abundance of some species, while others closely allied to them are very rare.

The Law of Population of Species.

The general proportion that must obtain between certain groups of animals is readily seen. Large animals cannot be so abundant as small ones; the carnivora must be less numerous than the herbivora; eagles and lions can never be so plentiful as pigeons and antelopes; and the wild asses of the Tartarian deserts cannot equal in numbers the horses of the more luxuriant prairies and pampas of America. The greater or less fecundity of an animal is often considered to be one of the chief causes of its abundance or scarcity; but a consideration of the facts will show us that it really has little or nothing to do with the matter. Even the least prolific of animals would increase rapidly if unchecked, whereas it is evident that the animal population of the globe must be stationary, or perhaps, through the influence of man, decreasing. Fluctuations there may be; but permanent increase, except in restricted localities, is almost impossible. For example, our own observation must convince us that birds do not go on increasing every year in a geometrical ratio, as they would do, were there not some powerful check to their natural increase. Very few birds produce less than two young ones each year, while many have six, eight, or ten; four will certainly be below the average; and if we suppose that each pair produce young only four times in their life, that will also be below the average, supposing them not to die either by violence or want of food. Yet at this rate how tremendous would be the increase in a few years from a single pair! A simple calculation will show that in fifteen years each pair of birds would have increased to nearly ten millions! * whereas we have no reason to believe that the number of the birds of any country increases at all in fifteen or in one hundred and fifty years. With such powers of in-

* This is under estimated. The number would really amount to more than two thousand millions!

crease the population must have reached its limits, and have become stationary, in a very few years after the origin of each species. It is evident, therefore, that each year an immense number of birds must perish—as many in fact as are born; and as on the lowest calculation the progeny are each year twice as numerous as their parents, it follows that, whatever be the average number of individuals existing in any given country, *twice that number must perish annually,*—a striking result, but one which seems at least highly probable, and is perhaps under rather than over the truth. It would therefore appear that, as far as the continuance of the species and the keeping up the average number of individuals are concerned, large broods are superfluous. On the average all above *one* become food for hawks and kites, wild cats or weasels, or perish of cold and hunger as winter comes on. This is strikingly proved by the case of particular species; for we find that their abundance in individuals bears no relation whatever to their fertility in producing offspring.

Perhaps the most remarkable instance of an immense bird population is that of the passenger pigeon of the United States, which lays only one, or at most two eggs, and is said to rear generally but one young one. Why is this bird so extraordinarily abundant, while others producing two or three times as many young are much less plentiful? The explanation is not difficult. The food most congenial to this species, and on which it thrives best, is abundantly distributed over a very extensive region, offering such differences of soil and climate, that in one part or another of the area the supply never fails. The bird is capable of a very rapid and long-continued flight, so that it can pass without fatigue over the whole of the district it inhabits, and as soon as the supply of food begins to fail in one place is able to discover a fresh feeding-ground. This example strikingly shows us that the procuring a constant supply of wholesome food is almost the sole condition requisite for ensuring the rapid increase of a given species, since neither the limited fecundity, nor the unrestrained attacks of birds of prey and of man are here sufficient to check it. In no other birds are these peculiar circumstances so strikingly combined. Either their food is more liable to failure, or they have not sufficient power of wing to search for it over an extensive area, or during some season of the year it becomes very scarce, and less wholesome substitutes have to be found ; and thus, though more fertile in offspring, they can never increase beyond the supply of food in the least favourable seasons.

Many birds can only exist by migrating, when their food becomes scarce, to regions possessing a milder, or at least a different climate, though, as these migrating birds are seldom excessively abundant, it is evident that the countries they visit are still deficient in a constant and abundant supply of wholesome food. Those whose organization does not permit them to migrate when their food becomes periodically scarce, can never attain a large population. This is probably the reasons why woodpeckers are scarce with us, while in the tropics they are among the most abundant of solitary birds. Thus the house sparrrow is more abundant than the redbreast, because its food is more constant and plentiful,—seeds of grasses being preserved during the winter, and our farm-yards and stubble-fields furnishing an almost inexhaustible supply. Why, as a general rule, are aquatic, and especially sea birds, very numerous in individuals ? Not because they are more prolific than others, generally the contrary ; but because their food never fails, the sea-shores and river-banks daily swarming with a fresh supply of small mollusca and crustacea. Exactly the same laws will apply to mammals. Wild cats are prolific and have few enemies ; why then are they never as abundant as rabbits ? The only intelligible answer is, that their supply of food is more precarious. It appears evident, therefore, that so long as a country remains physically unchanged, the numbers of its animal population cannot materially increase. If one species does so, some others requiring the same kind of food must diminish in proportion. The numbers that die annually must be immense ; and as the individual existence of each animal depends upon itself, those that die must be the weakest—the very young, the aged, and the diseased —while those that prolong their existence can only be the most perfect in health and vigour— those who are best able to obtain food regularly, and avoid their numerous enemies. It is, as we commenced by remarking, " a struggle for existence," in which the weakest and least perfectly organized must always succumb.

The Abundance or Rarity of a Species dependent upon its more or less perfect Adaptation to the Conditions of Existence.

It seems evident that what takes place among the individuals of a species must also occur among the several allied species of a group,—viz., that those which are best adapted to obtain a regular supply of food, and to defend themselves against the attacks of their enemies and the vicissitudes of the seasons, must necessarily obtain and preserve a superiority in population ; while those species which from some defect of power or organization are the least capable of counteracting the vicissitudes of food-supply, &c., must diminish in numbers, and, in extreme cases, become altogether extinct. Between these extremes the species will present various degrees of capacity for ensuring the means of preserving life ; and it is thus we account for the abundance or rarity of species. Our ignorance will generally prevent us from accurately tracing the effects to their causes ; but could we become perfectly acquainted with the

D

organization and habits of the various species of animals, and could we measure the capacity of each for performing the different acts necessary to its safety and existence under all the varying circumstances by which it is surrounded, we might be able even to calculate the proportionate abundance of individuals which is the necessary result.

If now we have succeeded in establishing these two points—1st, *that the animal population of a country is generally stationary, being kept down by a periodical deficiency of food, and other checks;* and, 2nd, *that the comparative abundance or scarcity of the individuals of the several species is entirely due to their organization and resulting habits, which, rendering it more difficult to procure a regular supply of food and to provide for their personal safety in some cases than in others, can only be balanced by a difference in the population which have to exist in a given area*—we shall be in a condition to proceed to the consideration of *varieties*, to which the preceding remarks have a direct and very important application.

Useful Variations will tend to Increase; useless or hurtful Variations to Diminish.

Most or perhaps all the variations from the typical form of a species must have some definite effect, however slight, on the habits or capacities of the individuals. Even a change of colour might, by rendering them more or less distinguishable, affect their safety; a greater or less development of hair might modify their habits. More important changes, such as an increase in the power or dimensions of the limbs or any of the external organs, would more or less affect their mode of procuring food or the range of country which they could inhabit. It is also evident that most changes would affect, either favourably or adversely, the powers of prolonging existence. An antelope with shorter or weaker legs must necessarily suffer more from the attacks of the feline carnivora; the passenger pigeon with less powerful wings would sooner or later be affected in its powers of procuring a regular supply of food; and in both cases the result must necessarily be a diminution of the population of the modified species. If, on the other hand, any species should produce a variety having slightly increased powers of preserving existence, that variety must inevitably in time acquire a superiority in numbers. These results must follow as surely as old age, intemperance, or scarcity of food produce an increased mortality. In both cases there may be many individual exceptions; but on the average the rule will invariably be found to hold good. All varieties will therefore fall into two classes—those which under the same conditions would never reach the population of the parent species, and those which would in time obtain and keep a numerical superiority. Now, let some alteration of physical conditions occur in the district—a long period of drought, a destruction of vegetation by locusts, the irruption of some new carnivorous animal seeking " pastures new "—any change in fact tending to render existence more difficult to the species in question, and tasking its utmost powers to avoid complete extermination ; it is evident that, of all the individuals composing the species, those forming the least numerous and most feebly organized variety would suffer first, and, were the pressure severe, must soon become extinct. The same causes continuing in action, the parent species would next suffer, would gradually diminish in numbers, and with a recurrence of similar unfavourable conditions might also become extinct. The superior variety would then alone remain, and on a return to favourable circumstances would rapidly increase in numbers and occupy the place of the extinct species and variety.

Superior Varieties will ultimately Extirpate the original Species.

The *variety* would now have replaced the *species*, of which it would be a more perfectly developed and more highly organized form. It would be in all respects better adapted to secure its safety, and to prolong its individual existence and that of the race. Such a variety *could not* return to the original form ; for that form is an inferior one, and could never compete with it for existence. Granted, therefore, a " tendency " to reproduce the original type of the species, still the variety must ever re-

main preponderant in numbers, and under adverse
physical conditions *again alone survive.* But this
new, improved, and populous race might itself, in
course of time, give rise to new varieties, exhibiting
several diverging modifications of form, any of which,
tending to increase the facilities for preserving existence, must, by the same general law, in their
turn become predominant. Here, then, we have
progression and continued divergence deduced from the
general laws which regulate the existence of animals
in a state of nature, and from the undisputed fact
that varieties do frequently occur. It is not, however, contended that this result would be invariable;
a change of physical conditions in the district might
at times materially modify it, rendering the race
which had been the most capable of supporting existence under the former conditions now the least so,
and even causing the extinction of the newer and,
for a time, superior race, while the old or parent
species and its first inferior varieties continued to
flourish. Variations in unimportant parts might
also occur, having no perceptible effect on the life-
preserving powers; and the varieties so furnished
might run a course parallel with the parent species,
either giving rise to further variations or returning
to the former type. All we argue for is, that certain varieties have a tendency to maintain their
existence longer than the original species, and this
tendency must make itself felt; for though the doctrine of chances or averages can never be trusted to

on a limited scale, yet, if applied to high numbers, the results come. nearer to what theory demands, and, as we approach to an infinity of examples, become strictly accurate. Now the scale on which nature works is so vast—the numbers of individuals and the periods of time with which she deals approach so near to infinity, than any cause, however slight, and however liable to be veiled and counteracted by accidental circumstances, must in the end produce its full legitimate results.

The Partial Reversion of Domesticated Varieties explained.

Let us now turn to domesticated animals, and inquire how varieties produced among them are affected by the principles here enunciated. The essential difference in the condition of wild and domestic animals is this,—that among the former, their well-being and very existence depend upon the full exercise and healthy condition of all their senses and physical powers, whereas, among the latter, these are only partially exercised, and in some cases are absolutely unused. A wild animal has to search, and often to labour, for every mouthful of food—to exercise sight, hearing, and smell in seeking it, and in avoiding dangers, in procuring shelter from the inclemency of the seasons, and in providing for the subsistence and safety of its offspring. There is no muscle of its body that is not called into daily and hourly activity; there is no sense or faculty that is not strengthened by continual exercise. The domestic animal, on the other hand, has food provided for it, is sheltered, and often confined, to guard it against the vicissitudes of the seasons, is carefully secured from the attacks of its natural enemies, and seldom even rears its young without human assistance. Half of its senses and faculties become quite useless, and the other half are but occasionally called into feeble exercise, while even its muscular system is only irregularly brought into action.

Now when a variety of such an animal occurs, having increased power or capacity in any organ or sense, such increase is totally useless, is never called into action, and may even exist without the animal ever becoming aware of it. In the wild animal, on the contrary, all its faculties and powers being brought into full action for the necessities of existence, any increase becomes immediately available, is strengthened by exercise, and must even slightly modify the food, the habits, and the whole economy of the race. It creates as it were a new animal, one of superior powers, and which will necessarily increase in numbers and outlive those which are inferior to it.

Again, in the domesticated animal all variations have an equal chance of continuance; and those which would decidedly render a wild animal unable to compete with its fellows and continue its existence are no disadvantage whatever in a state of domesticity. Our quickly fattening pigs, short-legged sheep pouter pigeons, and poodle dogs could never have come into existence in a state of nature, because the very first step towards such inferior forms would have led to the rapid extinction of the race; still less could they now exist in competition with their wild allies. The great speed but slight endurance of the race horse, the unwieldly strength of the ploughman's team, would both be useless in a state of nature. If turned wild on the pampas, such animals would probably soon become extinct, or under favourable circumstances might each gradually lose those extreme qualities which would never be called into action, and in a few generations revert to a common type, which must be that in which the various powers and faculties are so proportioned to each other as to be best adapted to procure food and secure safety,—that in which by the full exercise of every part of its organisation the animal can alone continue to live. Domestic varieties, when turned wild, *must* return to something near the type of the original wild stock, *or become altogether extinct.**

We see, then, that no inferences as to the permanence of varieties in a state of nature can be deduced from the observations of those occurring among domestic animals. The two are so much opposed to each other in every circumstance of their

---

* That is, they will vary, and the variations which tend to adapt them to the wild state, and therefore approximate them to wild animals, will be preserved. Those individuals which do not vary sufficiently will perish.

existence, that what applies to the one is almost sure not to apply to the other. Domestic animals are abnormal, irregular, artificial ; they are subject to variations which never occur and never can occur in a state of nature : their very existence depends altogether on human care; so far are many of them removed from that just proportion of faculties, that true balance of organisation, by means of which alone an animal left to its own resources can preserve its existence and continue its race.

Lamarck's Hypothesis very different from that now advanced.

The hypothesis of Lamarck—that progressive changes in species have been produced by the attempts of animals to increase the development of their own organs, and thus modify their structure and habits—has been repeatedly and easily refuted by all writers on the subject of varieties and species, and it seems to have been considered that when this was done the whole question has been finally settled ; but the view here developed renders such hypothesis quite unnecessary, by showing that similar results must be produced by the action of principles constantly at work in nature. The powerful retractile talons of the falcon- and the cat-tribes have not been produced or increased by the volition of those animals; but among the different varieties which occurred in the earlier and less highly organized forms of these groups, *those always survived longest which had the greatest facilities for seizing their prey.* Neither did the giraffe acquire its long neck by desiring to reach the foliage of the more lofty shrubs, and constantly stretching it neck for the purpose, but because any varieties which occurred among its antitypes with a longer neck than usual *at once secured a fresh range of pasture over the same ground as their shorter-necked companions, and on the first scarcity of food were thereby enabled to outlive them.* Even the peculiar colours of many animals, more especially of insects, so closely resembling the soil or leaves or bark on which they habitually reside, are explained on the same principle; for though in the course of ages varieties of many tints may have occurred, *yet those races having colours best adapted to concealment from their enemies would inevitably survive the longest.* We have also here an acting cause to account for that balance so often observed in nature,—a deficiency in one set of organs always being compensated by an increased development of some others—powerful wings accompanying weak feet, or great velocity making up for the absence of defensive weapons; for it has been shown that all varieties in which an unbalanced deficiency occurred could not long continue their existence. The action of this principle is exactly like that of the centrifugal governor of the steam engine, which checks and corrects any irregularities almost before they become evident; and in like manner no unbalanced deficiency in the animal kingdom can ever reach any conspicuous magnitude, because it would make itself felt at the very first step, by rendering existence difficult and extinction almost sure soon to follow. An origin such as is here advocated will also agree with the peculiar character of the modifications of form and structure which obtain in organized beings—the many lines of divergence from a central type, the increasing efficiency and power of a particular organ through a succession of allied species, and the remarkable persistence of unimportant parts, such as colour, texture of plumage and hair, form of horns or crests, through a series of species differing considerably in more essential characters. It also furnishes us with a reason for that " more specialized structure " which Professor Owen states to be a characteristic of recent compared with extinct forms, and which would evidently be the result of the progressive modification of any organ applied to a special purpose in the animal economy.

Conclusion.

We believe we have now shown that there is a tendency in nature to the continued progression of certain classes of *varieties* further and further from the original type—a progression to which there appears no reason to assign any definite limits—and that the same principle which produces this result in a state of nature will also explain why domestic varieties have a tendency, when they become wild, to revert to the original type. This progression, by minute steps, in various directions, but always checked and balanced by the necessary conditions, subject to which alone existence can be preserved, may, it is believed, be followed out so as to agree with all the phænomena presented by organized beings, their extinction and succession in past ages, and all the extraordinary modifications of form, instinct and habits which they exhibit.

III.

MIMICRY, AND OTHER PROTECTIVE RE-SEMBLANCES AMONG ANIMALS.

THERE is no more convincing proof of the truth of a comprehensive theory, than its power of absorbing and finding a place for new facts, and its capability of interpreting phænomena which had been previously looked upon as unaccountable anomalies. It is thus that the law of universal gravitation and the undulatory theory of light have become established and universally accepted by men of science. Fact after fact has been brought forward as being apparently inconsistent with them, and one after another these very facts have been shown to be the consequences of the laws they were at first supposed to disprove. A false theory will never stand this test. Advancing knowledge brings to light whole groups of facts which it cannot deal with, and its advocates steadily decrease in numbers, notwithstanding the ability and scientific skill with which it may have been supported. The great name of Edward Forbes did not prevent his theory of " Polarity in the distribution of Organic beings in Time " from dying a natural death; but the most striking illustration of the behaviour of a false theory is to be found in the " Circular and Quinarian System " of classification propounded by MacLeay, and developed by Swainson, with an amount of knowledge and ingenuity that have rarely been surpassed. This theory was eminently attractive, both from its symmetry and completeness, and from the interesting nature of the varied analogies and affinities which it brought to light and made use of. The series of Natural History volumes in " Lardner's Cabinet Cyclopædia," in which Mr. Swainson developed it in most departments of the animal kingdom, made it widely known ; and in fact for a long time these were the best and almost the only popular text-books for the rising generation of naturalists. It was favourably received too by the older school, which was perhaps rather an indication of its unsoundness. A considerable number of well-known naturalists either spoke approvingly of it, or advocated similar principles, and for a good many years it was decidedly in the ascendent. With such a favourable introduction, and with such talented exponents, it must have become established if it had had any germ of truth in it ; yet it quite died out in a few short years, its very existence is now a matter of history ; and so rapid was its fall that its talented creator, Swainson, perhaps lived to be the last man who believed in it.

Such is the course of a false theory. That of a true one is very different, as may be well seen by the progress of opinion on the subject of Natural Selection. In less than eight years "The Origin of Species " has produced conviction in the minds of a majority of the most eminent living men of science. New facts, new problems, new difficulties as they arise are accepted, solved or removed by this theory; and its principles are illustrated by the progress and conclusions of every well established branch of human knowledge. It is the object of the present essay to show how it has recently been applied to connect together and explain a variety of curious facts which had long been considered as inexplicable anomalies.

Importance of the Principle of Utility.

Perphaps no principle has ever been announced so fertile in results as that which Mr. Darwin so earnestly impresses upon us, and which is indeed a necessary deduction from the theory of Natural Selection, namely—that none of the definite facts of organic nature, no special organ, no characteristic form or marking, no peculiarities of instinct or of habit, no relations between species or between groups of species—can exist, but which must 'now be or once have been *useful* to the individuals or the races which possess them. This great principle gives us a clue which we can follow out in the study of many recondite phænomena, and leads us to seek a meaning and a purpose of some definite character in minutiæ which we should be otherwise almost sure to pass over as insignificant or unimportant.

Popular Theories of Colour in Animals.

The adaptation of the external colouring of animals

to their conditions of life has long been recognised, and has been imputed either to an originally created specific peculiarity, or to the direct action of climate, soil, or food. Where the former explanation has been accepted, it has completely checked inquiry, since we could never get any further than the fact of the adaptation. There was nothing more to be known about the matter. The second explanation was soon found to be quite inadequate to deal with all the varied phases of the phænomena, and to be contradicted by many well-known facts. For example, wild rabbits are always of grey or brown tints well suited for conceal-ment among grass and fern. But when these rabbits are domesticated, without any change of climate or food, they vary into white or black, and these varie-ties may be multiplied to any extent, forming white or black races. Exactly the same thing has occurred with pigeons; and in the case of rats and mice, the white variety has not been shown to be at all dependent on alteration of climate, food, or other external conditions. In many cases the wings of an insect not only assume the exact tint of the bark or leaf it is accustomed to rest on, but the form and veining of the leaf or the exact rugosity of the bark is imitated; and these detailed modifications cannot be reasonably imputed to climate or to food, since in many cases the species does not feed on the substance it resembles, and when it does, no reasonable connexion can be shown to exist between the supposed cause and the effect produced. It was reserved for the theory of Natural Selection to solve all these problems, and many others which were not at first supposed to be directly connected with them. To make these latter intelligible, it will be necessary to give a sketch of the whole series of phænomena which may be classed under the head of useful or protective resemblances.

Importance of Concealment as Influencing Colour.

Concealment, more or less complete, is useful to many animals, and absolutely essential to some. Those which have numerous enemies from which they cannot escape by rapidity of motion, find safety in concealment. Those which prey upon others must also be so constituted as not to alarm them by their presence or their approach, or they would soon die of hunger. Now it is remarkable in how many cases nature gives this boon to the animal, by colouring it with such tints as may best serve to enable it to escape from its enemies or to entrap its prey Desert animals as a rule are desert-coloured. The lion is a typical example of this, and must be almost invisible when crouched upon the sand or among desert rocks and stones. Antelopes are all more or less sandy-coloured. The camel is pre-eminently so. The Egyptian cat and the Pampas cat are sandy or earth-coloured. The Australian kangaroos are of the same tints, and the original colour of the wild horse is supposed to have been a sandy or clay-colour.

E

The desert birds are still more remarkably protected by their assimilative hues. The stonechats, the larks, the quails, the goatsuckers and the grouse, which abound in the North African and Asiatic deserts, are all tinted and mottled so as to resemble with wonderful accuracy the average colour and aspect of the soil in the district they inhabit. The Rev. H. Tristram, in his account of the ornithology of North Africa in the 1st volume of the "Ibis," says : "In the desert, where neither trees, brushwood, nor even undulation of the surface afford the slightest protection to its foes, a modification of colour which shall be assimilated to that of the surrounding country, is absolutely necessary. Hence *without exception* the upper plumage of *every bird,* whether lark, chat, sylvain, or sand-grouse, and also the fur of *all the smaller mammals,* and the skin of *all the snakes and lizards,* is of one uniform isabelline or sand colour." After the testimony of so able an observer it is unnecessary to adduce further examples of the protective colours of desert animals.

Almost equally striking are the cases of arctic animals possessing the white colour that best conceals them upon snowfields and icebergs. The polar bear is the only bear that is white, and it lives constantly among snow and ice. The arctic fox, the ermine and the alpine hare change to white in winter only, because in summer white would be more conspicuous than any other colour, and therefore a danger rather than a protection; but the

American polar hare, inhabiting regions of almost perpetual snow, is white all the year round. Other animals inhabiting the same Northern regions do not, however, change colour. The sable is a good example, for throughout the severity of a Siberian winter it retains its rich brown fur. But its habits are such that it does not need the protection of colour, for it is said to be able to subsist on fruits and berries in winter, and to be so active upon the trees as to catch small birds among the branches. So also the woodchuck of Canada has a dark-brown fur ; but then it lives in burrows and frequents river banks, catching fish and small animals that live in or near the water.

Among birds, the ptarmigan is a fine example of protective colouring. Its summer plumage so exactly harmonizes with the lichen-coloured stones among which it delights to sit, that a person may walk through a flock of them without seeing a single bird ; while in winter its white plumage is an almost equal protection. The snow-bunting, the jer-falcon, and the snowy owl are also white-coloured birds inhabiting the arctic regions, and there can be little doubt but that their colouring is to some extent protective.

Nocturnal animals supply us with equally good illustrations. Mice, rats, bats, and moles possess the least conspicuous of hues, and must be quite invisible at times when any light colour would be instantly seen. Owls and goatsuckers are of those dark mottled tints that will assimilate with bark and lichen, and thus protect them during the day, and at the same time be inconspicuous in the dusk.

It is only in the tropics, among forests which never lose their foliage, that we find whole groups of birds whose chief colour is green. The parrots are the most striking example, but we have also a group of green pigeons in the East; and the barbets, leaf-thrushes, bee-eaters, white-eyes, turacos, and several smaller groups, have so much green in their plumage as to tend greatly to conceal them among the foliage.

Special Modifications of Colour.

The conformity of tint which has been so far shown to exist between animals and their habitations is of a somewhat general character; we will now consider the cases of more special adaptation. If the lion is enabled by his sandy colour readily to conceal himself by merely crouching down upon the desert, how, it may be asked, do the elegant markings of the tiger, the jaguar, and the other large cats agree with this theory? We reply that these are generally cases of more or less special adaptation. The tiger is a jungle animal, and hides himself among tufts of grass or of bamboos, and in these positions the vertical stripes with which his body is adorned must so assimilate with the vertical stems of the bamboo, as to assist greatly in concealing him from his approaching prey. How remarkable it is that besides the lion and tiger, almost all the other large cats are arboreal in their habits, and almost all have ocellated or spotted skins, which must certainly tend to blend them with the background of foliage; while the one exception, the puma, has an ashy brown uniform fur, and has the habit of clinging so closely to a limb of a tree while waiting for his prey to pass beneath as to be hardly distinguishable from the bark.

Among birds, the ptarmigan, already mentioned, must be considered a remarkable case of special adaptation. Another is a South-American goatsucker (Caprimulgus rupestris) which rests in the bright sunshine on little bare rocky islets in the Upper Rio Negro, where its unusually light colours so closely resemble those of the rock and sand, that it can scarcely be detected till trodden upon.

The Duke of Argyll, in his "Reign of Law," has pointed out the admirable adaptation of the colours of the woodcock to its protection. The various browns and yellows and pale ash-colour that occur in fallen leaves are all reproduced in its plumage, so that when according to its habit it rests upon the ground under trees, it is almost impossible to detect it. In snipes the colours are modified so as to be equally in harmony with the prevalent forms and colours of marshy vegetation. Mr. J. M. Lester, in a paper read before the Rugby School Natural History Society, observes : — "The wood-dove, when perched amongst the branches of its favourite *fir*, is scarcely discernible; whereas, were it among some

lighter foliage, the blue and purple tints in its plumage
would far sooner betray it. The robin redbreast too,
although it might be thought that the red on its breast
made it much easier to be seen, is in reality not at
all endangered by it, since it generally contrives to
get among some russet or yellow fading leaves, where
the red matches very well with the autumn tints,
and the brown of the rest of the body with the bare
branches."

Reptiles offer us many similar examples. The most
arboreal lizards, the iguanas, are as green as the leaves
they feed upon, and the slender whip-snakes are rendered almost invisible as they glide among the foliage
by a similar colouration. How difficult it is sometimes to catch sight of the little green tree-frogs
sitting on the leaves of a small plant enclosed in a
glass case in the Zoological Gardens; yet how much
better concealed must they be among the fresh green
damp foliage of a marshy forest. There is a North-
American frog found on lichen-covered rocks and
walls, which is so coloured as exactly to resemble
them, and as long as it remains quiet would certainly
escape detection. Some of the geckos which cling
motionless on the trunks of trees in the tropics, are
of such curiously marbled colours as to match exactly
with the bark they rest upon.

In every part of the tropics there are tree-snakes
that twist among boughs and shrubs, or lie coiled up
on the dense masses of foliage. These are of many
distinct groups, and comprise both venomous and

harmless genera; but almost all of them are of a beautiful green colour, sometimes more or less adorned with white or dusky bands and spots. There can be little doubt that this colour is doubly useful to them, since it will tend to conceal them from their enemies, and will lead their prey to approach them unconscious of danger. Dr. Gunther informs me that there is only one genus of true arboreal snakes (Dipsas) whose colours are rarely green, but are of various shades of black, brown, and olive, and these are all nocturnal reptiles, and there can be little doubt conceal themselves during the day in holes, so that the green protective tint would be useless to them, and they accordingly retain the more usual reptilian hues.

Fishes present similar instances. Many flat fish, as for example the flounder and the skate, are exactly the colour of the gravel or sand on which they habitually rest. Among the marine flower gardens of an Eastern coral reef the fishes present every variety of gorgeous colour, while the river fish even of the tropics rarely if ever have gay or conspicuous markings. A very curious case of this kind of adaptation occurs in the sea-horses (Hippocampus) of Australia, some of which bear long foliaceous appendages resembling seaweed, and are of a brilliant red colour; and they are known to live among seaweed of the same hue, so that when at rest they must be quite invisible. There are now in the aquarium of the Zoological Society some slender green pipe - fish which fasten themselves to any object at the bottom by their prehensile tails, and float about with the current, looking exactly like some simple cylindrical algæ.

It is, however, in the insect world that this principle of the adaptation of animals to their environment is most fully and strikingly developed. In order to understand how general this is, it is necessary to enter somewhat into details, as we shall thereby be better able to appreciate the significance of the still more remarkable phenomena we shall presently have to discuss. It seems to be in proportion to their sluggish motions or the absence of other means of defence, that insects possess the protective colouring. In the tropics there are thousands of species of insects which rest during the day clinging to the bark of dead or fallen trees; and the greater portion of these are delicately mottled with gray and brown tints, which though symmetrically disposed and infinitely varied, yet blend so completely with the usual colours of the bark, that at two or three feet distance they are quite undistinguishable. In some cases a species is known to frequent only one species of tree. This is the case with the common South American long-horned beetle (Onychocerus scorpio) which, Mr. Bates informed me, is found only on a rough-barked tree, called Tapiribá, on the Amazon. It is very abundant, but so exactly does it resemble the bark in colour and rugosity, and so closely does it cling to the branches, that until it moves it is absolutely invisible! An allied species (O.

concentricus) is found only at Pará, on a distinct species of tree, the bark of which it resembles with equal accuracy. Both these insects are abundant, and we may fairly conclude that the protection they derive from this strange concealment is at least one of the causes that enable the race to flourish.

Many of the species of Cicindela, or tiger beetle, will illustrate this mode of protection. Our common Cicindela campestris frequents grassy banks, and is of a beautiful green colour, while C. maritima, which is found only on sandy sea-shores, is of a pale bronzy yellow, so as to be almost invisible. A great number of the species found by myself in the Malay islands are similarly protected. The beautiful Cicindela gloriosa, of a very deep velvety green colour, was only taken upon wet mossy stones in the bed of a mountain stream, where it was with the greatest difficulty detected. A large brown species (C. heros) was found chiefly on dead leaves in forest paths; and one which was never seen except on the wet mud of salt marshes was of a glossy olive so exactly the colour of the mud as only to be distinguished when the sun shone, by its shadow! Where the sandy beach was coralline and nearly white, I found a very pale Cicindela; wherever it was volcanic and black, a dark species of the same genus was sure to be met with.

There are in the East small beetles of the family Buprestidæ which generally rest on the midrib of a leaf, and the naturalist often hesitates before picking them off, so closely do they resemble pieces of bird's dung. Kirby and Spence mention the small beetle Onthophilus sulcatus as being like the seed of an umbelliferous plant; and another small weevil, which is much persecuted by predatory beetles of the genus Harpalus, is of the exact colour of loamy soil, and was found to be particularly abundant in loam pits. Mr. Bates mentions a small beetle (Chlamys pilula) which was undistinguishable by the eye from the dung of caterpillars, while some of the Cassidæ, from their hemispherical forms and pearly gold colour, resemble glittering dew-drops upon the leaves.

A number of our small brown and speckled weevils at the approach of any object roll off the leaf they are sitting on, at the same time drawing in their legs and antennæ, which fit so perfectly into cavities for their reception that the insect becomes a mere oval brownish lump, which it is hopeless to look for among the similarly coloured little stones and earth pellets among which it lies motionless.

The distribution of colour in butterflies and moths respectively is very instructive from this point of view. The former have all their brilliant colouring on the upper surface of all four wings, while the under surface is almost always soberly coloured, and often very dark and obscure. The moths on the contrary have generally their chief colour on the hind wings only, the upper wings being of dull, sombre, and often imitative tints, and these generally conceal the hind wings when the insects are in repose. This arrangement of the colours is therefore eminently protective, because the butterfly always rests with his wings raised so as to conceal the dangerous brilliancy of his upper surface. It is probable that if we watched their habits sufficiently we should find the under surface of the wings of butterflies very frequently imitative and protective. Mr. T. W. Wood has pointed out that the little orange-tip butterfly often rests in the evening on the green and white flower heads of an umbelliferous plant, and that when observed in this position the beautiful green and white mottling of the under surface completely assimilates with the flower heads and renders the creature very difficult to be seen. It is probable that the rich dark colouring of the under side of our peacock, tortoiseshell, and red-admiral butterflies answers a similar purpose.

Two curious South American butterflies that always settle on the trunks of trees (Gynecia dirce and Callizona acesta) have the under surface curiously striped and mottled, and when viewed obliquely must closely assimilate with the appearance of the furrowed bark of many kinds of trees. But the most wonderful and undoubted case of protective resemblance in a butterfly which I have ever seen, is that of the common Indian Kallima inachis, and its Malayan ally, Kallima paralekta. The upper surface of these insects is very striking and showy, as they are of a large size, and are adorned with a broad band of rich orange on a deep bluish ground. The under side is very variable in colour, so that out of fifty specimens no two can be found exactly alike, but every one of them will be of some shade of ash or brown or ochre, such as are found among dead, dry, or decaying leaves.   The apex of the upper wings is produced into an acute point, a very common form in the leaves of tropical shrubs and trees, and the lower wings are also produced into a short narrow tail.   Between these two points runs a dark curved line exactly representing the midrib of a leaf, and from this radiate on each side a few oblique lines, which serve to indicate the lateral veins of a leaf. These marks are more clearly seen on the outer portion of the base of the wings, and on the inner side towards the middle and apex, and it is very curious to observe how the usual marginal and transverse striæ of the group are here modified and strengthened so as to become adapted for an imitation of the venation of a leaf.   We come now to a still more extraordinary part of the imitation, for we find representations of leaves in every stage of decay, variously blotched and mildewed and pierced with holes, and in many cases irregularly covered with powdery black dots gathered into patches and spots, so closely resembling the various kinds of minute fungi that grow on dead leaves that it is impossible to avoid thinking at first sight that the butterflies themselves have been attacked by real fungi.

But this resemblance, close as it is, would be of little use if the habits of the insect did not accord with it. If the butterfly sat upon leaves or upon flowers, or opened its wings so as to expose the upper surface, or exposed and moved its head and antennæ as many other butterflies do, its disguise would be of little avail. We might be sure, however, from the analogy of many other cases, that the habits of the insect are such as still further to aid its deceptive garb; but we are not obliged to make any such supposition, since I myself had the good fortune to observe scores of Kallima paralekta, in Sumatra, and to capture many of them, and can vouch for the accuracy of the following details. These butterflies frequent dry forests and fly very swiftly. They were never seen to settle on a flower or a green leaf, but were many times lost sight of in a bush or tree of dead leaves. On such occasions they were generally searched for in vain, for while gazing intently at the very spot where one had disappeared, it would often suddenly dart out, and again vanish twenty or fifty yards further on. On one or two occasions the insect was detected reposing, and it could then be seen how completely it assimilates itself to the surrounding leaves. It sits on a nearly upright twig, the wings fitting closely back to back, concealing the antennæ and head, which are drawn up between their bases. The little tails of the hind wing touch the branch, and form a perfect stalk to the leaf, which is supported in its place by the claws of the middle pair of feet, which are slender and inconspicuous. The irregular outline of the wings gives exactly the perspective effect of a shrivelled leaf. We thus have size, colour, form, markings, and habits, all combining together to produce a disguise which may be said to be absolutely perfect; and the protection which it affords is sufficiently indicated by the abundance of the individuals that possess it.

The Rev. Joseph Greene has called attention to the striking harmony between the colours of those British moths which are on the wing in autumn and winter, and the prevailing tints of nature at those seasons. In autumn various shades of yellow and brown prevail, and he shows that out of fifty-two species that fly at this season, no less than forty-two are of corresponding colours. Orgyia antiqua, O. gonostigma, the genera Xanthia, Glæa, and Ennomos are examples. In winter, gray and silvery tints prevail, and the genus Chematobia and several species of Hybernia which fly during this season are of corresponding hues. No doubt if the habits of moths in a state of nature were more closely observed, we should find many cases of special protective resemblance. A few such have already been noticed. Agriopis aprilina, Acronycta psi, and many other moths which rest during the day on the north side of the trunks of trees can with difficulty be distinguished from the grey and green lichens that cover them. The lappet moth (Gastropacha querci) closely resembles both in shape and colour a brown dry leaf; and the well-known buff-tip moth, when at rest is like the broken end of a lichen-covered branch. There are some of the small moths which exactly resemble the dung of birds dropped on leaves, and on this point Mr. A. Sidgwick, in a paper read before the Rugby School Natural History Society, gives the following original observation :—" I myself have more than once mistaken Cilix compressa, a little white and grey moth, for a piece of bird's dung dropped upon a leaf, and *vice versâ* the dung for the moth. Bryophila Glandifera and Perla are the very image of the mortar walls on which they rest; and only this summer, in Switzerland, I amused myself for some time in watching a moth, probably Larentia tripunctaria, fluttering about quite close to me, and then alighting on a wall of the stone of the district which it so exactly matched as to be quite invisible a couple of yards off." There are probably hosts of these resemblances which have not been observed, owing to the difficulty of finding many of the species in their stations of natural repose. Caterpillars are also similarly protected. Many exactly resemble in tint the leaves they feed upon ; others are like little brown twigs, and many are so strangely marked or humped, that when motionless they can hardly be taken to be living creatures at all. Mr. Andrew Murray has remarked how closely the larva of the peacock moth (Saturnia pavonia-minor) harmonizes in its ground colour with that of the young buds of heather on which it feeds, and that the pink spots with which it is decorated correspond with the flowers and flower-buds of the same plant.

The whole order of Orthoptera, grasshoppers, locusts, crickets, &c., are protected by their colours harmonizing with that of the vegetation or the soil on which they live, and in no other group have we such striking examples of special resemblance. Most of the

tropical Mantidæ and Locustidæ are of the exact tint of
the leaves on which they habitually repose, and many
of them in addition have the veinings of their wings
modified so as exactly to imitate that of a leaf. This
is carried to the furthest possible extent in the wonderful genus, Phyllium, the "walking leaf," in which not
only are the wings perfect imitations of leaves in every
detail, but the thorax and legs are flat, dilated, and
leaf-like; so that when the living insect is resting
among the foliage on which it feeds, the closest observation is often unable to distinguish between the
animal and the vegetable.

The whole family of the Phasmidæ, or spectres, to
which this insect belongs, is more or less imitative, and
a great number of the species are called " walking-stick
insects," from their singular resemblance to twigs and
branches. Some of these are a foot long and as thick
as one's finger, and their whole colouring, form, rugosity, and the arrangement of the head, legs, and antennæ, are such as to render them absolutely identical in
appearance with dead sticks. They hang loosely about
shrubs in the forest, and have the extraordinary habit
of stretching out their legs unsymmetrically, so as to
render the deception more complete. One of these
creatures obtained by myself in Borneo (Ceroxylus
laceratus) was covered over with foliaceous excrescences
of a clear olive green colour, so as exactly to resemble a
stick grown over by a creeping moss or jungermannia.
The Dyak who brought it me assured me it was grown
over with moss although alive, and it was only after a

most minute examination that I could convince myself
it was not so.

We need not adduce any more examples to show
how important are the details of form and of colouring
in animals, and that their very existence may often
depend upon their being by these means concealed from
their enemies. This kind of protection is found apparently in every class and order, for it has been noticed
wherever we can obtain sufficient knowledge of the
details of an animal's life-history. It varies in degree,
from the mere absence of conspicuous colour or a
general harmony with the prevailing tints of nature,
up to such a minute and detailed resemblance to inorganic or vegetable structures as to realize the talisman
of the fairy tale, and to give its possessor the power of
rendering itself invisible.

Theory of Protective Colouring.

We will now endeavour to show how these wonderful
resemblances have most probably been brought about.
Returning to the higher animals, let us consider the
remarkable fact of the rarity of white colouring in the
mammalia or birds of the temperate or tropical zones
in a state of nature. There is not a single white land-
bird or quadruped in Europe, except the few arctic or
alpine species, to which white is a protective colour.
Yet in many of these creatures there seems to be no
inherent tendency to avoid white, for directly they are
domesticated white varieties arise, and appear to thrive
as well as others. We have white mice and rats, white

F

cats, horses, dogs, and cattle, white poultry, pigeons, turkeys, and ducks, and white rabbits. Some of these animals have been domesticated for a long period, others only for a few centuries; but in almost every case in which an animal has been thoroughly domesticated, parti-coloured and white varieties are produced and become permanent.

It is also well known that animals in a state of nature produce white varieties occasionally. Blackbirds, starlings, and crows are occasionally seen white, as well as elephants, deer, tigers, hares, moles, and many other animals; but in no case is a permanent white race produced. Now there are no statistics to show that the normal-coloured parents produce white offspring oftener under domestication than in a state of nature, and we have no right to make such an assumption if the facts can be accounted for without it. But if the colours of animals do really, in the various instances already adduced, serve for their concealment and preservation, then white or any other conspicuous colour must be hurtful, and must in most cases shorten an animal's life. A white rabbit would be more surely the prey of hawk or buzzard, and the white mole, or field mouse, could not long escape from the vigilant owl. So, also, any deviation from those tints best adapted to conceal a carnivorous animal would render the pursuit of its prey much more difficult, would place it at a disadvantage among its fellows, and in a time of scarcity would probably cause it to starve to death. On the other hand, if an animal spreads from a temperate into an arctic district, the conditions are changed. During a large portion of the year, and just when the struggle for existence is most severe, white is the prevailing tint of nature, and dark colours will be the most conspicuous. The white varieties will now have an advantage; they will escape from their enemies or will secure food, while their brown companions will be devoured or will starve; and as "like produces like" is the established rule in nature, the white race will become permanently established, and dark varieties, when they occasionally appear, will soon die out from their want of adaptation to their environment. In each case the fittest will survive, and a race will be eventually produced adapted to the conditions in which it lives.

We have here an illustration of the simple and effectual means by which animals are brought into harmony with the rest of nature. That slight amount of variability in every species, which we often look upon as something accidental or abnormal, or so insignificant as to be hardly worthy of notice, is yet the foundation of all those wonderful and harmonious resemblances which play such an important part in the economy of nature. Variation is generally very small in amount, but it is all that is required, because the change in the external conditions to which an animal is subject is generally very slow and intermittent. When these changes have taken place too rapidly, the result has often been the extinction of species; but the general rule is, that climatal and geological changes go on slowly, and the slight but continual variations in the colour, form, and structure of all animals, has furnished individuals adapted to these changes, and who have become the progenitors of modified races. Rapid multiplication, incessant slight variation, and survival of the fittest—these are the laws which ever keep the organic world in harmony with the inorganic, and with itself. These are the laws which we believe have produced all the cases of protective resemblance already adduced, as well as those still more curious examples we have yet to bring before our readers.

It must always be borne in mind that the more wonderful examples, in which there is not only a general but a special resemblance—as in the walking leaf, the mossy phasma, and the leaf-winged butterfly—represent those few instances in which the process of modification has been going on during an immense series of generations. They all occur in the tropics, where the conditions of existence are the most favourable, and where climatic changes have for long periods been hardly perceptible. In most of them favourable variations both of colour, form, structure, and instinct or habit, must have occurred to produce the perfect adaptation we now behold. All these are known to vary, and favourable variations when not accompanied by others that were unfavourable, would certainly survive. At one time a little step might be made in this direction, at another time in that—a change of conditions might sometimes render useless that which it had taken ages to produce—great and sudden physical modifications might often produce the extinction of a race just as it was approaching perfection, and a hundred checks of which we can know nothing may have retarded the progress towards perfect adaptation; so that we can hardly wonder at there being so few cases in which a completely successful result has been attained as shown by the abundance and wide diffusion of the creatures so protected.

Objection that Colour, as being dangerous, should not exist in Nature.

It is as well here to reply to an objection that will no doubt occur to many readers—that if protection is so useful to all animals, and so easily brought about by variation and survival of the fittest, there ought to be no conspicuously-coloured creatures; and they will perhaps ask how we account for the brilliant birds, and painted snakes, and gorgeous insects, that occur abundantly all over the world. It will be advisable to answer this question rather fully, in order that we may be prepared to understand the phenomena of "mimicry," which it is the special object of this paper to illustrate and explain.

The slightest observation of the life of animals will show us, that they escape from their enemies and obtain their food in an infinite number of ways; and that their varied habits and instincts are in every case adapted to the conditions of their existence. The porcupine and the hedgehog have a defensive armour that saves them from the attacks of most animals.

The tortoise is not injured by the conspicuous colours of his shell, because that shell is in most cases an effectual protection to him.   The skunks of North America find safety in their power of emitting an unbearably offensive odour; the beaver in its aquatic habits and solidly constructed abode.   In some cases the chief danger to an animal occurs at one particular period of its existence, and if that is guarded against its numbers can easily be maintained.   This is the case with many birds, the eggs and young of which are especially obnoxious to danger, and we find accordingly a variety of curious contrivances to protect them.   We have nests carefully concealed, hung from the slender extremities of grass or boughs over water, or placed in the hollow of a tree with a very small opening.   When these precautions are successful, so many more individuals will be reared than can possibly find food during the least favourable seasons, that there will always be a number of weakly and inexperienced young birds who will fall a prey to the enemies of the race, and thus render necessary for the stronger and healthier individuals no other safeguard than their strength and activity.   The instincts most favourable to the production and rearing of offspring will in these cases be most important, and the survival of the fittest will act so as to keep up and advance those instincts, while other causes which tend to modify colour and marking may continue their action almost unchecked.

It is perhaps in insects that we may best study the varied means by which animals are defended or concealed. One of the uses of the phosphorescence with which many insects are furnished, is probably to frighten away their enemies; for Kirby and Spence state that a ground beetle (Carabus) has been observed running round and round a luminous centipede as if afraid to attack it. An immense number of insects have stings, and some stingless ants of the genus Polyrachis are armed with strong and sharp spines on the back, which must render them unpalatable to many of the smaller insectivorous birds. Many beetles of the family Curculionidæ have the wing cases and other external parts so excessively hard, that they cannot be pinned without first drilling a hole to receive the pin, and it is probable that all such find a protection in this excessive hardness. Great numbers of insects hide themselves among the petals of flowers, or in the cracks of bark and timber; and finally, extensive groups and even whole orders have a more or less powerful and disgusting smell and taste, which they either possess permanently, or can emit at pleasure. The attitudes of some insects may also protect them, as the habit of turning up the tail by the harmless rove-beetles (Staphylindidæ) no doubt leads other animals besides children to the belief that they can sting. The curious attitude assumed by sphinx caterpillars is probably a safeguard, as well as the blood-red tentacles which can suddenly be thrown out from the neck, by the caterpillars of all the true swallow-tailed butterflies.

It is among the groups that possess some of these varied kinds of protection in a high degree, that we find the greatest amount of conspicuous colour, or at least the most complete absence of protective imitation. The stinging Hymenoptera, wasps, bees, and hornets, are, as a rule, very showy and brilliant insects, and there is not a single instance recorded in which any one of them is coloured so as to resemble a vegetable or inanimate substance. The Chrysididæ, or golden wasps, which do not sting, possess as a substitute the power of rolling themselves up into a ball, which is almost as hard and polished as if really made of metal,—and they are all adorned with the most gorgeous colours. The whole order Hemiptera (comprising the bugs) emit a powerful odour, and they present a very large proportion of gay-coloured and conspicuous insects. The lady-birds (Coccinellidæ) and their allies the Eumorphidæ, are often brightly spotted, as if to attract attention; but they can both emit fluids of a very disagreeable nature, they are certainly rejected by some birds, and are probably never eaten by any.

The great family of ground beetles (Carabidæ) almost all possess a disagreeable and some a very pungent smell, and a few, called bombardier beetles, have the peculiar faculty of emitting a jet of very volatile liquid, which appears like a puff of smoke, and is accompanied by a distinct crepitating explosion. It is probably because these insects are mostly nocturnal and predacious that they do not present more vivid hues. They are chiefly remarkable for brilliant metallic tints or dull red patches when they are not wholly black, and are therefore very conspicuous by day, when insect-eaters are kept off by their bad odour and taste, but are sufficiently invisible at night when it is of importance that their prey should not become aware of their proximity.

It seems probable that in some cases that which would appear at first to be a source of danger to its possessor may really be a means of protection. Many showy and weak-flying butterflies have a very broad expanse of wing, as in the brilliant blue Morphos of Brazilian forests, and the large Eastern Papilios; yet these groups are tolerably plentiful. Now, specimens of these butterflies are often captured with pierced and broken wings, as if they had been seized by birds from whom they had escaped; but if the wings had been much smaller in proportion to the body, it seems probable that the insect would be more frequently struck or pierced in a vital part, and thus the increased expanse of the wings may have been indirectly beneficial.

In other cases the capacity of increase in a species is so great that however many of the perfect insect may be destroyed, there is always ample means for the continuance of the race. Many of the flesh flies, gnats, ants, palm-tree weevils and locusts are in this category. The whole family of Cetoniadæ or rose chafers, so full of gaily-coloured species, are probably saved from attack by a combination of characters. They fly very rapidly with a zigzag or waving course; they hide themselves the moment they alight, either in the corolla of flowers, or in rotten wood, or in cracks and hollows of trees, and they are generally encased in a very hard and polished coat of mail which may render them unsatisfactory food to such birds as would be able to capture them. The causes which lead to the development of colour have been here able to act unchecked, and we see the result in a large variety of the most gorgeously-coloured insects.

Here, then, with our very imperfect knowledge of the life-history of animals, we are able to see that there are widely varied modes by which they may obtain protection from their enemies or concealment from their prey. Some of these seem to be so complete and effectual as to answer all the wants of the race, and lead to the maintenance of the largest possible population. When this is the case, we can well understand that no further protection derived from a modification of colour can be of the slightest use, and the most brilliant hues may be developed without any prejudicial effect upon the species. On some of the laws that determine the development of colour something may be said presently. It is now merely necessary to show that concealment by obscure or imitative tints is only one out of very many ways by which animals maintain their existence; and having done this we are prepared to consider the phenomena of what has been termed " mimicry." It is to be particularly observed, however, that the word is not here used in the sense of voluntary imitation, but to imply a particular kind of resemblance—a resemblance not in internal structure but in external appearance—a resemblance in those parts only that catch the eye—a resemblance that deceives. As this kind of resemblance has the same effect as voluntary imitation or mimicry, and as we have no word that expresses the required meaning, "mimicry" was adopted by Mr. Bates (who was the first to explain the facts), and has led to some misunderstanding ; but there need be none, if it is remembered that both "mimicry" and "imitation" are used in a metaphorical sense, as implying that close external likeness which causes things unlike in structure to be mistaken for each other.

Mimicry.

It has been long known to entomologists that certain insects bear a strange external resemblance to others belonging to distinct genera, families, or even orders, and with which they have no real affinity whatever. The fact, however, appears to have been generally considered as dependent upon some unknown law of "analogy"—some "system of nature," or "general plan," which had guided the Creator in designing the myriads of insect forms, and which we could never hope to understand. In only one case does it appear that the resemblance was thought to be useful, and to have been designed as a means to a definite and intelligible purpose. The flies of the genus Volucella enter the nests of bees to deposit their eggs, so that their larvæ may feed upon the larvæ of the bees, and these flies are each wonderfully like the bee on which it is parasitic. Kirby and Spence believed that this resemblance or "mimicry" was for the express purpose of protecting the flies from the attacks of the bees, and the connection is so evident that it was hardly possible to avoid this conclusion. The resemblance, however, of moths to butterflies or to bees, of beetles to wasps, and of locusts to beetles, has been many times noticed by eminent writers ; but scarcely ever till within the last few years does it appear to have been considered that these resemblances had any special purpose, or were of any direct benefit to the insects themselves. In this respect they were looked upon as accidental, as instances of the " curious analogies " in nature which must be wondered at but which could not be explained. Recently, however, these instances have been greatly multiplied; the nature of the resemblances has been more carefully studied, and it has been found that they are often carried out into such details as almost to imply a purpose of deceiving the observer. The phenomena, moreover, have been shown to follow certain definite laws, which again all indicate their dependence on the more general law of the " survival of the fittest," or " the preservation of favoured races in the struggle for life." It will, perhaps, be as well here to state what these laws or general conclusions are, and then to give some account of the facts which support them.

The first law is, that in an overwhelming majority of cases of mimicry, the animals (or the groups) which resemble each other inhabit the same country, the same district, and in most cases are to be found together on the very same spot.

The second law is, that these resemblances are not indiscriminate, but are limited to certain groups, which in every case are abundant in species and individuals, and can often be ascertained to have some special protection.

The third law is, that the species which resemble or " mimic " these dominant groups, are comparatively less abundant in individuals, and are often very rare.

These laws will be found to hold good, in all the cases of true mimicry among various classes of animals to which we have now to call the attention of our readers.

Mimicry among Lepidoptera.

As it is among butterflies that instances of mimicry are most numerous and most striking, an account of some of the more prominent examples in this group will first be given. There is in South America an extensive family of these insects, the Heliconidæ, which are in many respects very remarkable. They are so abundant and characteristic in all the woody portions of the American tropics, that in almost every locality they will be seen more frequently than any other butterflies. They are distinguished by very elongate wings, body, and antennæ, and are exceedingly beautiful and varied in their colours ; spots and patches of yellow, red, or pure white upon a black, blue, or brown ground, being most general. They frequent the forests chiefly, and all fly slowly and weakly; yet although they are so conspicuous, and could certainly be caught by insectivorous birds more easily than almost any other insects, their great abundance all over the wide region they inhabit shows that they are not so persecuted. It is to be especially remarked also, that they possess no adaptive colouring to protect them during repose, for the under side of their wings presents the same, or at least an equally conspicuous colouring as the upper side; and they may be observed after sunset suspended at the end of twigs and leaves where they have taken up their station for the night, fully exposed to the attacks of enemies if they have any. These beautiful insects possess, however, a strong pungent semi-aromatic or medicinal odour, which seems to pervade all the juices of their system. When the entomologist squeezes the breast of one of them between his fingers to kill it, a yellow liquid exudes which stains the skin, and the smell of which can only be got rid of by time and repeated washings. Here we have probably the cause of their immunity from attack, since there is a great deal of evidence to show that certain insects are so disgusting to birds that they will under no circumstances touch them. Mr. Stainton has observed that a brood of young turkeys greedily devoured all the worthless moths he had amassed in a night's "sugaring," yet one after another seized and rejected a single white moth which happened to be among them. Young pheasants and partridges which eat many kinds of caterpillars seem to have an absolute dread of that of the common currant moth, which they will never touch, and tomtits as well as other small birds appear never to eat the same species. In the case of the Heliconidæ, however, we have some direct evidence to the same effect. In the Brazilian forests there are great numbers of insectivorous birds—as jacamars, trogons, and puffbirds —which catch insects on the wing, and that they destroy many butterflies is indicated by the fact that the wings of these insects are often found on the ground where their bodies have been devoured. But among these there are no wings of Heliconidæ, while those of the large showy Nymphalidæ, which have a much swifter flight, are often met with. Again, a gentleman who had recently returned from Brazil stated at a meeting of the Entomological Society that he once observed a pair of puffbirds catching butterflies, which they brought to their nest to feed their young; yet during half an hour they never brought one of the Heliconidæ, which were flying lazily about in great numbers, and which they could have captured more easily than any others. It was this circumstance that led Mr. Belt to observe them so long, as he could not understand why the most common insects should be altogether passed by. Mr. Bates also tells us that he never saw them molested by lizards or predacious flies, which often pounce on other butterflies.

If, therefore, we accept it as highly probable (if not proved) that the Heliconidæ are very greatly protected from attack by their peculiar odour and taste, we find it much more easy to understand their chief characteristics —their great abundance, their slow flight, their gaudy colours, and the entire absence of protective tints on

their under surfaces. This property places them somewhat in the position of those curious wingless birds of
oceanic islands, the dodo, the apteryx, and the moas,
which are with great reason supposed to have lost the
power of flight on account of the absence of carnivorous
quadrupeds. Our butterflies have been protected in a
different way, but quite as effectually; and the result
has been that as there has been nothing to escape from,
there has been no weeding out of slow flyers, and as
there has been nothing to hide from, there has been no
extermination of the bright-coloured varieties, and no
preservation of such as tended to assimilate with surrounding objects.

Now let us consider how this kind of protection must
act. Tropical insectivorous birds very frequently sit on
dead branches of a lofty tree, or on those which overhang
forest paths, gazing intently around, and darting off at
intervals to seize an insect at a considerable distance,
which they generally return to their station to devour.
If a bird began by capturing the slow-flying, conspicuous
Heliconidæ, and found them always so disagreeable that
it could not eat them, it would after a very few trials
leave off catching them at all; and their whole appearance, form, colouring, and mode of flight is so peculiar,
that there can be little doubt birds would soon learn to
distinguish them at a long distance, and never waste
any time in pursuit of them. Under these circumstances,
it is evident that any other butterfly of a group which
birds were accustomed to devour, would be almost
equally well protected by closely resembling a Heliconia

externally, as if it acquired also the disagreeable odour; always supposing that there were only a few of them among a great number of the Heliconias. If the birds could not distinguish the two kinds externally, and there were on the average only one eatable among fifty uneatable, they would soon give up seeking for the eatable ones, even if they knew them to exist. If, on the other hand, any particular butterfly of an eatable group acquired the disagreeable taste of the Heliconias while it retained the characteristic form and colouring of its own group, this would be really of no use to it whatever; for the birds would go on catching it among its eatable allies (compared with which it would rarely occur), it would be wounded and disabled, even if rejected, and its increase would thus be as effectually checked as if it were devoured. It is important, therefore, to understand that if any one genus of an extensive family of eatable butterflies were in danger of extermination from insect-eating birds, and if two kinds of variation were going on among them, some individuals possessing a slightly disagreeable taste, others a slight resemblance to the Heliconidæ, this latter quality would be much more valuable than the former. The change in flavour would not at all prevent the variety from being captured as before, and it would almost certainly be thoroughly disabled before being rejected. The approach in colour and form to the Heliconidæ, however, would be at the very first a positive, though perhaps a slight advantage; for although at short distances this variety would be easily distinguished and devoured, yet at a longer distance it might be mistaken for one of the uneatable group, and so be passed by and gain another day's life, which might in many cases be sufficient for it to lay a quantity of eggs and leave a numerous progeny, many of which would inherit the peculiarity which had been the safeguard of their parent.

Now, this hypothetical case is exactly realized in South America. Among the white butterflies forming the family Pieridæ (many of which do not greatly differ in appearance from our own cabbage butterflies) is a genus of rather small size (Leptalis), some species of which are white like their allies, while the larger number exactly resemble the Heliconidæ in the form and colouring of the wings. It must always be remembered that these two families are as absolutely distinguished from each other by structural characters as are the carnivora and the ruminants among quadrupeds, and that an entomologist can always distinguish the one from the other by the structure of the feet, just as certainly as a zoologist can tell a bear from a buffalo by the skull or by a tooth. Yet the resemblance of a species of the one family to another species in the other family was often so great, that both Mr. Bates and myself were many times deceived at the time of capture, and did not discover the distinctness of the two insects till a closer examination detected their essential differences. During his residence of eleven years in the Amazon valley, Mr. Bates found a number of species or varieties of Leptalis, each of which was a more or less exact copy of one of the Heliconidæ of the district it inhabited; and the results of his observations are embodied in a paper published in the Linnean Transactions, in which he first explained the phenomena of " mimicry" as the result of natural selection, and showed its identity in cause and purpose with protective resemblance to vegetable or inorganic forms.

The imitation of the Heliconidæ by the Leptalides is carried out to a wonderful degree in form as well as in colouring. The wings have become elongated to the same extent, and the antennæ and abdomen have both become lengthened, to correspond with the unusual condition in which they exist in the former family. In colouration there are several types in the different genera of Heliconidæ. The genus Mechanitis is generally of a rich semi-transparent brown, banded with black and yellow ; Methona is of large size, the wings transparent like horn, and with black transverse bands ; while the delicate Ithomias are all more or less transparent, with black veins and borders, and often with marginal and transverse bands of orange red. These different forms are all copied by the various species of Leptalis, every band and spot and tint of colour, and the various degrees of transparency, being exactly reproduced. As if to derive all the benefit possible from this protective mimicry, the habits have become so modified that the Leptalides generally frequent the very same spots as their models, and have the same mode of flight; and as they are always very scarce (Mr. Bates estimating their numbers at about one to a thousand of the group they resemble), there is hardly a possibility of their being found out by their enemies. It is also very remarkable that in almost every case the particular Ithomias and other species of Heliconidæ which they resemble, are noted as being very common species, swarming in individuals, and found over a wide range of country. This indicates antiquity and permanence in the species, and is exactly the condition most essential both to aid in the development of the resemblance, and to increase its utility.

But the Leptalides are not the only insects who have prolonged their existence by imitating the great protected group of Heliconidæ;—a genus of quite another family of most lovely small American butterflies, the Erycinidæ, and three genera of diurnal moths, also present species which often mimic the same dominant forms, so that some, as Ithomia ilerdina of St. Paulo, for instance, have flying with them a few individuals of three widely different insects, which are yet disguised with exactly the same form, colour, and markings, so as to be quite undistinguishable when upon the wing. Again, the Heliconidæ are not the only group that are imitated, although they are the most frequent models. The black and red group of South American Papilios, and the handsome Erycinian genus Stalachtis, have also a few who copy them; but this fact offers no difficulty, since these two groups are almost as dominant as the Heliconidæ. They both fly very slowly, they are both conspicuously coloured, and they both abound in individuals; so that there is every reason to believe that they possess a protection of a similar kind to the Heliconidæ, and that it is therefore equally an advantage to other insects to be mistaken for them. There is also another extraordinary fact that we are not yet in a position clearly to comprehend : some groups of the Heliconidæ themselves mimic other groups. Species of Heliconia mimic Mechanitis, and every species of Napeogenes mimics some other Heliconideous butterfly. This would seem to indicate that the distasteful secretion is not produced alike by all members of the family, and that where it is deficient protective imitation comes into play. It is this, perhaps, that has caused such a general resemblance among the Heliconidæ, such a uniformity of type with great diversity of colouring, since any aberration causing an insect to cease to look like one of the family would inevitably lead to its being attacked, wounded, and exterminated, even although it was not eatable.

In other parts of the world an exactly parallel series of facts have been observed. The Danaidæ and the Acræidæ of the Old World tropics form in fact one great group with the Heliconidæ. They have the same general form, structure, and habits : they possess the same protective odour, and are equally abundant in individuals, although not so varied in colour, blue and white spots on a black ground being the most general pattern. The insects which mimic these are chiefly Papilios, and Diadema, a genus allied to our peacock and tortoiseshell butterflies. In tropical Africa there is a peculiar group of the genus Danais, characterized by dark-brown and bluish-white colours, arranged in bands or stripes. One of these, Danais niavius, is exactly imitated both by Papilio hippocoon and by Diadema anthedon; another, Danais echeria, by Papilio cenea; and in Natal a variety of the Danais is found having a white spot at the tip of wings, accompanied by a variety of the Papilio bearing a corresponding white spot. Acræa gea is copied in its very peculiar style of colouration by the female of Papilio cynorta, by Panopæa hirce, and by the female of Elymnias phegea. Acræa euryta of Calabar has a female variety of Panopea hirce from the same place which exactly copies it; and Mr. Trimen, in his paper on Mimetic Analogies among African Butterflies, published in the Transactions of the Linnæan Society for 1868, gives a list of no less than sixteen species and varieties of Diadema and its allies, and ten of Papilio, which in their colour and markings are perfect mimics of species or varieties of Danais or Acræa which inhabit the same districts.

Passing on to India, we have Danais tytia, a butterfly with semi-transparent bluish wings and a border of rich reddish brown. This remarkable style of colouring is exactly reproduced in Papilio agestor and in Diadema nama, and all three insects not unfrequently come together in collections made at Darjeeling. In the Philippine Islands the large and curious Idea leuconöe with its semi-transparent white wings, veined and spotted with black, is copied by the rare Papilio idæoides from the same islands.

In the Malay archipelago the very common and

beautiful Euplæa midamus is so exactly mimicked by two rare Papilios (P. paradoxa and P. ænigma) that I generally caught them under the impression that they were the more common species; and the equally common and even more beautiful Euplæa rhadamanthus, with its pure white bands and spots on a ground of glossy blue and black, is reproduced in the Papilio caunus. Here also there are species of Diadema imitating the same group in two or three instances; but we shall have to adduce these further on in connexion with another branch of the subject.

It has been already mentioned that in South America there is a group of Papilios which have all the characteristics of a protected race, and whose peculiar colours and markings are imitated by other butterflies not so protected. There is just such a group also in the East, having very similar colours and the same habits, and these also are mimicked by other species in the same genus not closely allied to them, and also by a few of other families. Papilio hector, a common Indian butterfly of a rich black colour spotted with crimson, is so closely copied by Papilio romulus, that the latter insect has been thought to be its female. A close examination shows, however, that it is essentially different, and belongs to another section of the genus. Papilio antiphus and P. diphilus, black swallow-tailed butterflies with cream-coloured spots, are so well imitated by varieties of P. theseus, that several writers have classed them as the same species. Papilio liris, found only in the island of Timor, is accompanied

there by P. ænomaus, the female of which so exactly
resembles it that they can hardly be separated in the
cabinet, and on the wing are quite undistinguishable.
But one of the most curious cases is the fine yellow-
spotted Papilio cöon, which is unmistakeably imitated
by the female tailed form of Papilio memnon. These
are both from Sumatra; but in North India P. cöon
is replaced by another species, which has been named
P. doubledayi, having red spots instead of yellow;
and in the same district the corresponding female
tailed form of Papilio androgeus, sometimes considered
a variety of P. memnon, is similarly red-spotted. Mr.
Westwood has described some curious day-flying moths
(Epicopeia) from North India, which have the form
and colour of Papilios of this section, and two of these
are very good imitations of Papilio polydorus and
Papilio varuna, also from North India.

Almost all these cases of mimicry are from the
tropics, where the forms of life are more abundant,
and where insect development especially is of unchecked
luxuriance; but there are also one or two instances in
temperate regions. In North America, the large and
handsome red and black butterfly Danais erippus is
very common; and the same country is inhabited by
Limenitis archippus, which closely resembles the
Danais, while it differs entirely from every species of
its own genus.

The only case of probable mimicry in our own country is the following :—A very common white moth
(Spilosoma menthastri) was found by Mr. Stainton

to be rejected by young turkeys among hundreds of other moths on which they greedily fed. Each bird in succession took hold of this moth and threw it down again, as if too nasty to eat. Mr. Jenner Weir also found that this moth was refused by the Bullfinch, Chaffinch, Yellow Hammer, and Red Bunting, but eaten after much hesitation by the Robin. We may therefore fairly conclude that this species would be disagreeable to many other birds, and would thus have an immunity from attack, which may be the cause of its great abundance and of its conspicuous white colour. Now it is a curious thing that there is another moth, Diaphora mendica, which appears about the same time, and whose female only is white. It is about the same size as Spilosoma menthastri, and sufficiently resembles it in the dusk, and this moth is much less common. It seems very probable, therefore, that these species stand in the same relation to each other as the mimicking butterflies of various families do to the Heliconidæ and Danaidæ. It would be very interesting to experiment on all white moths, to ascertain if those which are most common are generally rejected by birds. It may be anticipated that they would be so, because white is the most conspicuous of all colours for nocturnal insects, and had they not some other protection would certainly be very injurious to them.

Lepidoptera mimicking other Insects.

In the preceding cases we have found Lepidoptera imitating other species of the same order, and such species only as we have good reason to believe were free from the attacks of many insectivorous creatures; but there are other instances in which they altogether lose the external appearance of the order to which they belong, and take on the dress of bees or wasps—insects which have an undeniable protection in their stings. The Sesiidæ and Ægeriidæ, two families of day-flying moths, are particularly remarkable in this respect, and a mere inspection of the names given to the various species shows how the resemblance has struck everyone. We have apiformis, vespiforme, ichneumoniforme, scoliæforme, sphegiforme (bee-like, wasp-like, ichneumon-like, &c.) and many others, all indicating a resemblance to stinging Hymenoptera. In Britain we may particularly notice Sesia bombiliformis, which very closely resembles the male of the large and common humble bee, Bombus hortorum ; Sphecia craboniforme, which is coloured like a hornet, and is (on the authority of Mr. Jenner Weir) much more like it when alive than when in the cabinet, from the way in which it carries its wings ; and the currant clear-wing, Trochilium tipuliforme, which resembles a small black wasp (Odynerus sinuatus) very abundant in gardens at the same season. It has been so much the practice to look upon these resemblances as mere curious analogies playing no part in the economy of nature, that we have scarcely any observations of the habits and appearance when alive of the hundreds of species of these groups in various parts of the world, or how far they are accompanied by Hymenoptera, which they specifically resemble. There are many species in India (like those figured by Professor Westwood in his "Oriental Entomology") which have the hind legs very broad and densely hairy, so as exactly to imitate the brush-legged bees (Scopulipedes) which abound in the same country. In this case we have more than mere resemblance of colour, for that which is an important functional structure in the one group is imitated in another whose habits render it perfectly useless.

Mimicry among Beetles.

It may fairly be expected that if these imitations of one creature by another really serve as a protection to weak and decaying species, instances of the same kind will be found among other groups than the Lepidoptera ; and such is the case, although they are seldom so prominent and so easily recognised as those already pointed out as occurring in that order. A few very interesting examples may, however, be pointed out in most of the other orders of insects. The Coleoptera or beetles that imitate other Coleoptera of distinct groups are very numerous in tropical countries, and they generally follow the laws already laid down as regulating these phenomena. The insects which others imitate always have a special protection, which leads them to be avoided as dangerous or uneatable by small insectivorous animals ; some have a disgusting taste (analogous to that of the Heliconidæ) ; others have such a hard and stony covering that they cannot be crushed or digested ; while a third set are very active,

and armed with powerful jaws, as well as having some
disagreeable secretion. Some species of Eumorphidæ
and Hispidæ, small flat or hemispherical beetles which
are exceedingly abundant, and have a disagreeable secretion, are imitated by others of the very distinct
group of Longicornes (of which our common musk-
beetle may be taken as an example). The extraordinary little Cyclopeplus batesii, belongs to the same
sub-family of this group as the Onychocerus scorpio
and O. concentricus, which have already been adduced
as imitating with such wonderful accuracy the bark
of the trees they habitually frequent; but it differs
totally in outward appearance from every one of its
allies, having taken upon itself the exact shape and
colouring of a globular Corynomalus, a little stinking
beetle with clubbed antennæ. It is curious to see how
these clubbed antennæ are imitated by an insect belonging to a group with long slender antennæ. The
sub-family Anisocerinæ, to which Cyclopeplus belongs,
is characterised by all its members possessing a little
knob or dilatation about the middle of the antennæ.
This knob is considerably enlarged in C. batesii, and
the terminal portion of the antennæ beyond it is so
small and slender as to be scarcely visible, and thus an
excellent substitute is obtained for the short clubbed
antennæ of the Corynomalus. Erythroplatis corallifer
is another curious broad flat beetle, that no one would
take for a Longicorn, since it almost exactly resembles
Cephalodonta spinipes, one of the commonest of the
South American Hispidæ; and what is still more

remarkable, another Longicorn of a distinct group, Streptolabis hispoides, was found by Mr. Bates, which resembles the same insect with equal minuteness,—a case exactly parallel to that among butterflies, where species of two or three distinct groups mimicked the same Heliconia. Many of the soft-winged beetles (Malacoderms) are excessively abundant in individuals, and it is probable that they have some similar protection, more especially as other species often strikingly resemble them. A Longicorn beetle, Pæciloderma terminale, found in Jamaica, is coloured exactly in the same way as a Lycus (one of the Malacoderms) from the same island. Eroschema poweri, a Longicorn from Australia, might certainly be taken for one of the same group, and several species from the Malay Islands are equally deceptive. In the Island of Celebes I found one of this group, having the whole body and elytra of a rich deep blue colour, with the head only orange ; and in company with it an insect of a totally different family (Eucnemidæ) with identically the same colouration, and of so nearly the same size and form as to completely puzzle the collector on every fresh occasion of capturing them. I have been recently informed by Mr. Jenner Weir, who keeps a variety of small birds, that none of them will touch our common " soldiers and sailors " (species of Malacoderms), thus confirming my belief that they were a protected group, founded on the fact of their being at once very abundant, of conspicuous colours, and the objects of mimicry.

There are a number of the larger tropical weevils which have the elytra and the whole covering of the body so hard as to be a great annoyance to the entomologist, because in attempting to transfix them the points of his pins are constantly turned. I have found it necessary in these cases to drill a hole very carefully with the point of a sharp penknife before attempting to insert a pin. Many of the fine long-antennæd Anthribidæ (an allied group) have to be treated in the same way. We can easily understand that after small birds have in vain attempted to eat these insects, they should get to know them by sight, and ever after leave them alone, and it will then be an advantage for other insects which are comparatively soft and eatable, to be mistaken for them. We need not be surprised, therefore, to find that there are many Longicorns which strikingly resemble the "hard beetles" of their own district. In South Brazil, Acanthotritus dorsalis is strikingly like a Curculio of the hard genus Heiliplus, and Mr. Bates assures me that he found Gymnocerus cratosomoides (a Longicorn) on the same tree with a hard Cratosomus (a weevil), which it exactly mimics. Again, the pretty Longicorn, Phacellocera batesii, mimics one of the hard Anthribidæ of the genus Ptychoderes, having long slender antennæ. In the Moluccas we find Cacia anthriboides, a small Longicorn which might be easily mistaken for a very common species of Anthribidæ found in the same districts; and the very rare Capnolymma stygium closely imitates the common Mecocerus gazella, which abounded where it was taken. Doliops curculionoides and other allied

Longicorns from the Philippine Islands most curiously resemble, both in form and colouring, the brilliant Pachyrhynchi,—Curculionidæ, which are almost peculiar to that group of islands. The remaining family of Coleoptera most frequently imitated is the Cicindelidæ. The rare and curious Longicorn, Collyrodes lacordairei, has exactly the form and colouring of the genus Collyris, while an undescribed species of Heteromera is exactly like a Therates, and was taken running on the trunks of trees, as is the habit of that group. There is one curious example of a Longicorn mimicking a Longicorn, like the Papilios and Heliconidæ which mimic their own allies. Agnia fasciata, belonging to the sub-family Hypselominæ, and Nemophas grayi, belonging to the Lamiinæ, were taken in Amboyna on the same fallen tree at the same time, and were supposed to be the same species till they were more carefully examined, and found to be structurally quite different. The colouring of these insects is very remarkable, being rich steel-blue black, crossed by broad hairy bands of orange buff, and out of the many thousands of known species of Longicorns they are probably the only two which are so coloured. The Nemophas grayi is the larger, stronger, and better armed insect, and belongs to a more widely spread and dominant group, very rich in species and individuals, and is therefore most probably the subject of mimicry by the other species.

Beetles mimicking other Insects.

We will now adduce a few cases in which beetles

imitate other insects, and insects of other orders imitate beetles.

Charis melipona, a South American Longicorn of the family Necydalidæ, has been so named from its resemblance to a small bee of the genus Melipona. It is one of the most remarkable cases of mimicry, since the beetle has the thorax and body densely hairy like the bee, and the legs are tufted in a manner most unusual in the order Coleoptera. Another Longicorn, Odontocera odyneroides, has the abdomen banded with yellow, and constricted at the base, and is altogether so exactly like a small common wasp of the genus Odynerus, that Mr. Bates informs us he was afraid to take it out of his net with his fingers for fear of being stung. Had Mr. Bates's taste for insects been less omnivorous than it was, the beetle's disguise might have saved it from his pin, as it had no doubt often done from the beak of hungry birds. A larger insect, Sphecomorpha chalybea, is exactly like one of the large metallic blue wasps, and like them has the abdomen connected with the thorax by a pedicel, rendering the deception most complete and striking. Many Eastern species of Longicorns of the genus Oberea, when on the wing exactly resemble Tenthredinidæ, and many of the small species of Hesthesis run about on timber, and cannot be distinguished from ants. There is one genus of South American Longicorns that appears to mimic the shielded bugs of the genus Scutellera. The Gymnocerous capucinus is one of these, and is very like Pachyotris fabricii, one of the Scutelleridæ. The beautiful Gymnocerous dulcissimus is also very like the same group of insects, though there is no known species that exactly corresponds to it; but this is not to be wondered at, as the tropical Hemiptera have been comparatively so little cared for by collectors.

Insects mimicking Species of other Orders.

The most remarkable case of an insect of another order mimicking a beetle is that of the Condylodera tricondyloides, one of the cricket family from the Philippine Islands, which is so exactly like a Tricondyla (one of the tiger beetles), that such an experienced entomologist as Professor Westwood placed it among them in his cabinet, and retained it there a long time before he discovered his mistake ! Both insects run along the trunks of trees, and whereas Tricondylas are very plentiful, the insect that mimics it is, as in all other cases, very rare. Mr. Bates also informs us that he found at Santarem on the Amazon, a species of locust which mimicked one of the tiger beetles of the genus Odontocheila, and was found on the same trees which they frequented.

There are a considerable number of Diptera, or two-winged flies, that closely resemble wasps and bees, and no doubt derive much benefit from the wholesome dread which those insects excite. The Midas dives, and other species of large Brazilian flies, have dark wings and metallic blue elongate bodies, resembling the large stinging Sphegidæ of the same country; and a very large fly of the genus Asilus has

H

black-banded wings and the abdomen tipped with rich orange, so as exactly to resemble the fine bee Euglossa dimidiata, and both are found in the same parts of South America. We have also in our own country species of Bombylius which are almost exactly like bees. In these cases the end gained by the mimicry is no doubt freedom from attack, but it has sometimes an altogether different purpose. There are a number of parasitic flies whose larvæ feed upon the larvæ of bees, such as the British genus Volucella and many of the tropical Bombylii, and most of these are exactly like the particular species of bee they prey upon, so that they can enter their nests unsuspected to deposit their eggs. There are also bees that mimic bees. The cuckoo bees of the genus Nomada are parasitic on the Andrenidæ, and they resemble either wasps or species of Andrena ; and the parasitic humble-bees of the genus Apathus almost exactly resemble the species of humble-bees in whose nests they are reared. Mr. Bates informs us that he found numbers of these " cuckoo" bees and flies on the Amazon, which all wore the livery of working bees peculiar to the same country.

There is a genus of small spiders in the tropics which feed on ants, and they are exactly like ants themselves, which no doubt gives them more opportunity of seizing their prey ; and Mr. Bates found on the Amazon a species of Mantis which exactly resembled the white ants which it fed upon, as well as several species of crickets (Scaphura), which resembled in a wonderful manner different sand-wasps of large size, which are constantly on the search for crickets with which to provision their nests.

Perhaps the most wonderful case of all is the large caterpillar mentioned by Mr. Bates, which startled him by its close resemblance to a small snake. The first three segments behind the head were dilatable at the will of the insect, and had on each side a large black pupillated spot, which resembled the eye of the reptile. Moreover, it resembled a poisonous viper, not a harmless species of snake, as was proved by the imitation of keeled scales on the crown produced by the recumbent feet, as the caterpillar threw itself backward!

The attitudes of many of the tropical spiders are most extraordinary and deceptive, but little attention has been paid to them. They often mimic other insects, and some, Mr. Bates assures us, are exactly like flower buds, and take their station in the axils of leaves, where they remain motionless waiting for their prey.

Cases of Mimicry among the Vertebrata.

Having thus shown how varied and extraordinary are the modes in which mimicry occurs among insects, we have now to enquire if anything of the same kind is to be observed among vertebrated animals. When we consider all the conditions necessary to produce a good deceptive imitation, we shall see at once that such can very rarely occur in the higher animals, since they possess none of those facilities for the almost infinite modifications of external form which exist in the very nature of insect organization. The outer covering of

insects being more or less solid and horny, they are
capable of almost any amount of change of form
and appearance without any essential modification
internally. In many groups the wings give much of
the character, and these organs may be much modified
both in form and colour without interfering with their
special functions. Again, the number of species of
insects is so great, and there is such diversity of form
and proportion in every group, that the chances of an
accidental approximation in size, form, and colour, of
one insect to another of a different group, are very
considerable ; and it is these chance approximations
that furnish the basis of mimicry, to be continually
advanced and perfected by the survival of those
varieties only which tend in the right direction.

In the Vertebrata, on the contrary, the skeleton
being internal the external form depends almost entirely on the proportions and arrangement of that
skeleton, which again is strictly adapted to the functions necessary for the well-being of the animal. The
form cannot therefore be rapidly modified by variation,
and the thin and flexible integument will not admit
of the development of such strange protuberances as
occur continually in insects. The number of species of
each group in the same country is also comparatively
small, and thus the chances of that first accidental
resemblance which is necessary for natural selection
to work upon are much diminished. We can hardly
see the possibility of a mimicry by which the elk could
escape from the wolf, or the buffalo from the tiger.

There is, however, in one group of Vertebrata such a general similarity of form, that a very slight modification, if accompanied by identity of colour, would produce the necessary amount of resemblance; and at the same time there exist a number of species which it would be advantageous for others to resemble, since they are armed with the most fatal weapons of offence. We accordingly find that reptiles furnish us with a very remarkable and instructive case of true mimicry.

Mimicry among Snakes.

There are in tropical America a number of venomous snakes of the genus Elaps, which are ornamented with brilliant colours disposed in a peculiar manner. The ground colour is generally bright red, on which are black bands of various widths and sometimes divided into two or three by yellow rings. Now, in the same country are found several genera of harmless snakes, having no affinity whatever with the above, but coloured exactly the same. For example, the poisonous Elaps fulvius often occurs in Guatemala with simple black bands on a coral-red ground; and in the same country is found the harmless snake Pliocerus equalis, coloured and banded in identically the same manner. A variety of Elaps corallinus has the black bands narrowly bordered with yellow on the same red ground colour, and a harmless snake, Homalocranium semi-cinctum, has exactly the same markings, and both are found in Mexico. The deadly Elaps lemniscatus has the black bands very broad, and each of them divided into three by narrow yellow rings; and this again is exactly copied by a harmless snake, Pliocerus elapoides, which is found along with its model in Mexico.

But, more remarkable still, there is in South America a third group of snakes, the genus Oxyrhopus, doubtfully venomous, and having no immediate affinity with either of the preceding, which has also the same curious distribution of colours, namely, variously disposed rings of red, yellow, and black; and there are some cases in which species of all three of these groups similarly marked inhabit the same district. For example, Elaps mipartitus has single black rings very close together. It inhabits the west side of the Andes, and in the same districts occur Pliocerus euryzonus and Oxyrhopus petolarius, which exactly copy its pattern. In Brazil Elaps lemniscatus is copied by Oxyrhopus trigeminus, both having black rings disposed in threes. In Elaps hemiprichii the ground colour appears to be black, with alternations of two narrow yellow bands and a broader red one; and of this pattern again we have an exact double in Oxyrhopus formosus, both being found in many localities of tropical South America.

What adds much to the extraordinary character of these resemblances is the fact, that nowhere in the world but in America are there any snakes at all which have this style of colouring. Dr. Gunther, of the British Museum, who has kindly furnished some of the details here referred to, assures me that this is the case; and that red, black, and yellow rings occur together on no other snakes in the world but on Elaps and the species which so closely resemble it. In all these cases, the size and form as well as the colouration, are so much alike, that none but a naturalist would distinguish the harmless from the poisonous species.

Many of the small tree-frogs are no doubt also mimickers. When seen in their natural attitudes, I have been often unable to distinguish them from beetles or other insects sitting upon leaves, but regret to say I neglected to observe what species or groups they most resembled, and the subject does not yet seem to have attracted the attention of naturalists abroad.

Mimicry among Birds.

In the class of birds there are a number of cases that make some approach to mimicry, such as the resemblance of the cuckoos, a weak and defenceless group of birds, to hawks and Gallinaceæ. There is, however, one example which goes much further than this, and seems to be of exactly the same nature as the many cases of insect mimicry which have been already given. In Australia and the Moluccas there is a genus of honeysuckers called Tropidorhynchus, good sized birds, very strong and active, having powerful grasping claws and long, curved, sharp beaks. They assemble together in groups and small flocks, and they have a very loud bawling note, which can be heard at a great distance, and serves to collect a number together in time of danger. They are very plentiful and very pugnacious, frequently driving away crows, and even hawks, which perch on a tree where a few of them are assembled. They are all of rather dull and obscure colours. Now in the same countries there is a group of orioles, forming the genus Mimeta, much weaker birds, which have lost the gay colouring of their allies the golden orioles, being usually olive-green or brown; and in several cases these most curiously resemble the Tropidorhynchus of the same island. For example, in the island of Bouru is found the Tropido-rhynchus bouruensis, of a dull earthy colour, and the Mimeta bouruensis, which resembles it in the follow-ing particulars :—The upper and under surfaces of the two birds are exactly of the same tints of dark and light brown; the Tropidorhynchus has a large bare black patch round the eyes; this is copied in the Mimeta by a patch of black feathers. The top of the head of the Tropidorhynchus has a scaly appearance from the narrow scale-formed feathers, which are imi-tated by the broader feathers of the Mimeta having a dusky line down each. The Tropidorhynchus has a pale ruff formed of curious recurved feathers on the nape (which has given the whole genus the name of Friar birds) ; this is represented in the Mimeta by a pale band in the same position. Lastly, the bill of the Tropidorhynchus is raised into a protuberant keel at the base, and the Mimeta has the same character, although it is not a common one in the genus. The result is, that on a superficial examination the birds are identical, although they have important structural differences,

and cannot be placed near each other in any natural
arrangement. As a proof that the resemblance is really
deceptive, it may be mentioned that the Mimeta is
figured and described as a honeysucker in the costly
"Voyage de l'Astrolabe," under the name of Philedon
bouruensis !

Passing to the island of Ceram, we find allied species
of both genera. The Tropidorhynchus subcornutus is
of an earthy brown colour washed with yellow ochre,
with bare orbits, dusky cheeks, and the usual pale recurved nape-ruff. The Mimeta forsteni is absolutely
identical in the tints of every part of the body, the
details of which are imitated in the same manner as
in the Bouru birds already described. In two other
islands there is an approximation towards mimicry,
although it is not so perfect as in the two preceding cases. In Timor the Tropidorhynchus timoriensis
is of the usual earthy brown above, with the nape-ruff
very prominent, the cheeks black, the throat nearly
white, and the whole under surface pale whitish brown.
These various tints are all well reproduced in Mimeta
virescens, the chief want of exact imitation being that
the throat and breast of the Tropidorhynchus has a
very scaly appearance, being covered with rigid pointed
feathers which are not imitated in the Mimeta, although
there are signs of faint dusky spots which may easily
furnish the groundwork of a more exact imitation by
the continued survival of favourable variations in the
same direction. There is also a large knob at the base
of the bill of the Tropidorhynchus which is not at all

imitated by the Mimeta. In the island of Morty (north of Gilolo) there exists the Tropidorhynchus fuscicapillus, of a dark sooty brown colour, especially on the head, while the under parts are rather lighter, and the characteristic ruff of the nape is wanting. Now it is curious that in the adjacent island of Gilolo should be found the Mimeta phæochromus, the upper surface of which is of exactly the same dark sooty tint as the Tropidorhynchus, and is the only known species that is of such a dark colour. The under side is not quite light enough, but it is a good approximation. This Mimeta is a rare bird, and may very probably exist in Morty, though not yet found there; or, on the other hand, recent changes in physical geography may have led to the restriction of the Tropidorhynchus to that island, where it is very common.

Here, then, we have two cases of perfect mimicry and two others of good approximation, occurring between species of the same two genera of birds; and in three of these cases the pairs that resemble each other are found together in the same island, and to which they are peculiar. In all these cases the Tropidorhynchus is rather larger than the Mimeta, but the difference is not beyond the limits of variation in species, and the two genera are somewhat alike in form and proportion. There are, no doubt, some special enemies by which many small birds are attacked, but which are afraid of the Tropidorhynchus (probably some of the hawks), and thus it becomes advantageous for the weak Mimeta to resemble the strong, pugnacious, noisy, and very abundant Tropidorhynchus.

My friend, Mr. Osbert Salvin, has given me another interesting case of bird mimicry. In the neighbourhood of Rio Janeiro is found an insect-eating hawk (Harpagus diodon), and in the same district a bird-eating hawk (Accipiter pileatus) which closely resembles it. Both are of the same ashy tint beneath, with the thighs and under wing-coverts reddish brown, so that when on the wing and seen from below they are undistinguishable. The curious point, however, is that the Accipiter has a much wider range than the Harpagus, and in the regions where the insect-eating species is not found it no longer resembles it, the under wing-coverts varying to white; thus indicating that the red-brown colour is kept true by its being useful to the Accipiter to be mistaken for the insect-eating species, which birds have learnt not to be afraid of.

Mimicry among Mammals.

Among the Mammalia the only case which may be true mimicry is that of the insectivorous genus Cladobates, found in the Malay countries, several species of which very closely resemble squirrels. The size is about the same, the long bushy tail is carried in the same way, and the colours are very similar. In this case the use of the resemblance must be to enable the Cladobates to approach the insects or small birds on which it feeds, under the disguise of the harmless fruit-eating squirrel.

Objections to Mr. Bates' Theory of Mimicry.

Having now completed our survey of the most prominent and remarkable cases of mimicry that have yet
been noticed, we must say something of the objections
that have been made to the theory of their production
given by Mr. Bates, and which we have endeavoured to
illustrate and enforce in the preceding pages. Three
counter explanations have been proposed. Professor
Westwood admits the fact of the mimicry and its probable use to the insect, but maintains that each species
was created a mimic for the purpose of the protection
thus afforded it. Mr. Andrew Murray, in his paper on
the " Disguises of Nature," inclines to the opinion that
similar conditions of food and of surrounding circumstances have acted in some unknown way to produce the
resemblances ; and when the subject was discussed before
the Entomological Society of London, a third objection
was added—that heredity or the reversion to ancestral
types of form and colouration, might have produced
many of the cases of mimicry.

Against the special creation of mimicking species
there are all the objections and difficulties in the way
of special creation in other cases, with the addition of
a few that are peculiar to it. The most obvious is,
that we have gradations of mimicry and of protective
resemblance—a fact which is strongly suggestive of a
natural process having been at work. Another very
serious objection is, that as mimicry has been shown
to be useful only to those species and groups which

are rare and probably dying out, and would cease to have any effect should the proportionate abundance of the two species be reversed, it follows that on the special-creation theory the one species must have been created plentiful, the other rare ; and, notwithstanding the many causes that continually tend to alter the proportions of species, these two species must have always been specially maintained at their respective proportions, or the very purpose for which they each received their peculiar characteristics would have completely failed. A third difficulty is, that although it is very easy to understand how mimicry may be brought about by variation and the survival of the fittest, it seems a very strange thing for a Creator to protect an animal by making it imitate another, when the very assumption of a Creator implies his power to create it so as to require no such circuitous protection. These appear to be fatal objections to the application of the special-creation theory to this particular case.

The other two supposed explanations, which may be shortly expressed as the theories of " similar conditions " and of " heredity," agree in making mimicry, where it exists, an adventitious circumstance not necessarily connected with the well-being of the mimicking species. But several of the most striking and most constant facts which have been adduced, directly contradict both these hypotheses. The law that mimicry is confined to a few groups only is one of these, for " similar conditions " must act more or less on all groups in a limited region, and " heredity " must influence all groups related to each other in an equal degree. Again, the general fact that those species which mimic others are rare, while those which are imitated are abundant, is in no way explained by either of these theories, any more than is the frequent occurrence of some palpable mode of protection in the imitated species. " Reversion to an ancestral type " no way explains why the imitator and the imitated always inhabit the very same district, whereas allied forms of every degree of nearness and remoteness generally inhabit different countries, and often different quarters of the globe; and neither it, nor " similar conditions," will account for the likeness between species of distinct groups being superficial only— a disguise, not a true resemblance; for the imitation of bark, of leaves, of sticks, of dung; for the resemblance between species in different orders, and even different classes and sub-kingdoms; and finally, for the graduated series of the phenomena, beginning with a general harmony and adaptation of tint in autumn and winter moths and in arctic and desert animals, and ending with those complete cases of detailed mimicry which not only deceive predacious animals, but puzzle the most experienced insect collectors and the most learned entomologists.

Mimicry by Female Insects only.

But there is yet another series of phenomena connected with this subject, which considerably strengthens the view here adopted, while it seems quite incompatible with either of the other hypotheses; namely, the relation of protective colouring and mimicry to the sexual differences of animals. It will be clear to every one that if two animals, which as regards " external conditions" and " hereditary descent," are exactly alike, yet differ remarkably in colouration, one resembling a protected species and the other not, the resemblance that exists in one only can hardly be imputed to the influence of external conditions or as the effect of heredity. And if, further, it can be proved that the one requires protection more than the other, and that in several cases it is that one which mimics the protected species, while the one that least requires protection never does so, it will afford very strong corroborative evidence that there is a real connexion between the necessity for protection and the phenomenon of mimicry. Now the sexes of insects offer us a test of the nature here indicated, and appear to furnish one of the most conclusive arguments in favour of the theory that the phenomena termed " mimicry " are produced by natural selection.

The comparative importance of the sexes varies much in different classes of animals. In the higher vertebrates, where the number of young produced at a birth is small and the same individuals breed many years in succession, the preservation of both sexes is almost equally important. In all the numerous cases in which the male protects the female and her offspring, or helps to supply them with food, his importance in the economy of nature is proportionately increased, though it is never perhaps quite equal to that of the female. In insects the case is very different; they pair but once in their lives, and the prolonged existence of the male is in most cases quite unnecessary for the continuance of the race. The female, however, must continue to exist long enough to deposit her eggs in a place adapted for the development and growth of the progeny. Hence there is a wide difference in the need for protection in the two sexes ; and we should, therefore, expect to find that in some cases the special protection given to the female was in the male less in amount or altogether wanting. The facts entirely confirm this expectation. In the spectre insects (Phasmidæ) it is often the females alone that so strikingly resemble leaves, while the males show only a rude approximation. The male Diadema misippus is a very handsome and conspicuous butterfly, without a sign of protective or imitative colouring, while the female is entirely unlike her partner, and is one of the most wonderful cases of mimicry on record, resembling most accurately the common Danais chrysippus, in whose company it is often found. So in several species of South American Pieris, the males are white and black, of a similar type of colouring to our own "cabbage" butterflies, while the females are rich yellow and buff, spotted and marked so as exactly to resemble species of Heliconidæ with which they associate in the forest. In the Malay archipelago is found a Diadema which had always been considered a male insect on account of its glossy metallic-blue tints, while its companion of sober brown was looked upon as the female. I discovered, however, that the reverse is the case, and that the rich and glossy colours of the female are imitative and protective, since they cause her exactly to resemble the common Euploea midamus of the same regions, a species which has been already mentioned in this essay as mimicked by another butterfly, Papilio paradoxa. I have since named this interesting species Diadema anomala (see the Transactions of the Entomological Society, 1869, p. 285). In this case, and in that of Diadema misippus, there is no difference in the habits of the two sexes, which fly in similar localities; so that the influence of "external conditions" cannot be invoked here as it has been in the case of the South American Pieris pyrrha and allies, where the white males frequent open sunny places, while the Heliconia-like females haunt the shades of the forest.

We may impute to the same general cause (the greater need of protection for the female, owing to her weaker flight, greater exposure to attack, and supreme importance)—the fact of the colours of female insects being so very generally duller and less conspicuous than those of the other sex. And that it is chiefly due to this cause rather than to what Mr. Darwin terms "sexual selection" appears to be shown by the otherwise inexplicable fact, that in the groups which have a protection of any kind independent of concealment, sexual differences of colour are either quite wanting or slightly developed. The

Heliconidæ and Danaidæ, protected by a disagreeable flavour, have the females as bright and conspicuous as the males, and very rarely differing at all from them. The stinging Hymenoptera have the two sexes equally well coloured. The Carabidæ, the Coccinellidæ, Chrysomelidæ, and the Telephori have both sexes equally conspicuous, and seldom differing in colours. The brilliant Curculios, which are protected by their hardness, are brilliant in both sexes. Lastly, the glittering Cetoniadæ and Buprestidæ, which seem to be protected by their hard and polished coats, their rapid motions, and peculiar habits, present few sexual differences of colour, while sexual selection has often manifested itself by structural differences, such as horns, spines, or other processes.

Cause of the dull Colours of Female Birds.

The same law manifests itself in Birds. The female while sitting on her eggs requires protection by concealment to a much greater extent than the male ; and we accordingly find that in a large majority of the cases in which the male birds are distinguished by unusual brilliancy of plumage, the females are much more obscure, and often remarkably plain-coloured. The exceptions are such as eminently to prove the rule, for in most cases we can see a very good reason for them. In particular, there are a few instances among wading and gallinaceous birds in which the female has decidedly more brilliant colours than the male ; but it is a most curious and interesting fact

RESEMBLANCES AMONG ANIMALS.

that in most if not all these cases the males sit upon the eggs; so that this exception to the usual rule almost demonstrates that it is because the process of incubation is at once very important and very dangerous, that the protection of obscure colouring is developed. The most striking example is that of the gray phalarope (Phalaropus fulicarius). When in winter plumage, the sexes of this bird are alike in colouration, but in summer the female is much the most conspicuous, having a black head, dark wings, and reddish-brown back, while the male is nearly uniform brown, with dusky spots. Mr. Gould in his " Birds of Great Britain " figures the two sexes in both winter and summer plumage, and remarks on the strange peculiarity of the usual colours of the two sexes being reversed, and also on the still more curious fact that the " male alone sits on the eggs," which are deposited on the bare ground. In another British bird, the dotterell, the female is also larger and more brightly-coloured than the male; and it seems to be proved that the males assist in incubation even if they do not perform it entirely, for Mr. Gould tells us, " that they have been shot with the breast bare of feathers, caused by sitting on the eggs." The small quail-like birds forming the genus Turnix have also generally large and bright-coloured females, and we are told by Mr. Jerdon in his " Birds of India " that " the natives report that during the breeding season the females desert their eggs and associate in flocks while the males are employed in hatching the eggs."

It is also an ascertained fact, that the females are more bold and pugnacious than the males. A further confimation of this view is to be found in the fact (not hitherto noticed) that in a large majority of the cases in which bright colours exist in both sexes incubation takes place in a dark hole or in a dome-shaped nest. Female kingfishers are often equally brilliant with the male, and they build in holes in banks. Bee-eaters, trogons, motmots, and toucans, all build in holes, and in none is there any difference in the sexes, although they are, without exception, showy birds. Parrots build in holes in trees, and in the majority of cases they present no marked sexual difference tending to concealment of the female. Woodpeckers are in the same category, since though the sexes often differ in colour, the female is not generally less conspicuous than the male. Wagtails and titmice build concealed nests, and the females are nearly as gay as their mates. The female of the pretty Australian bird Pardalotus punctatus, is very conspicuously spotted on the upper surface, and it builds in a hole in the ground. The gay-coloured hang-nests (Icterinæ) and the equally brilliant tanagers may be well contrasted; for the former, concealed in their covered nests, present little or no sexual difference of colour—while the open-nested tanagers have the females dull-coloured and sometimes with almost protective tints. No doubt there are many individual exceptions to the rule here indicated, because many and various causes have combined to determine both the colouration and the habits of birds. These have no doubt acted and re-acted on each other; and when conditions have changed one of these characters may often have become modified, while the other, though useless, may continue by hereditary descent an apparent exception to what otherwise seems a very general rule. The facts presented by the sexual differences of colour in birds and their mode of nesting, are on the whole in perfect harmony with that law of protective adaptation of colour and form, which appears to have checked to some extent the powerful action of sexual selection, and to have materially influenced the colouring of female birds, as it has undoubtedly done that of female insects.

Use of the gaudy Colours of many Caterpillars.

Since this essay was first published a very curious difficulty has been cleared up by the application of the general principle of protective colouring. Great numbers of caterpillars are so brilliantly marked and coloured as to be very conspicuous even at a considerable distance, and it has been noticed that such caterpillars seldom hide themselves. Other species, however, are green or brown, closely resembling the colours of the substances on which they feed, while others again imitate sticks, and stretch themselves out motionless from a twig so as to look like one of its branches. Now, as caterpillars form so large a part of the food of birds, it was not easy to understand why any of them should have such bright colours and mark ings as to make them specially visible. Mr. Darwin had put the case to me as a difficulty from another point of view, for he had arrived at the conclusion that brilliant colouration in the animal kingdom is mainly due to sexual selection, and this could not have acted in the case of sexless larvæ. Applying here the analogy of other insects, I reasoned, that since some caterpillars were evidently protected by their imitative colouring, and others by their spiny or hairy bodies, the bright colours of the rest must also be in some way useful to them. I further thought that as some butterflies and moths were greedily eaten by birds while others were distasteful to them, and these latter were mostly of conspicuous colours, so probably these brilliantly coloured caterpillars were distasteful, and therefore never eaten by birds. Distastefulness alone would however be of little service to caterpillars, because their soft and juicy bodies are so delicate, that if seized and afterwards rejected by a bird they would almost certainly be killed. Some constant and easily perceived signal was therefore necessary to serve as a warning to birds never to touch these uneatable kinds, and a very gaudy and conspicuous colouring with the habit of fully exposing themselves to view becomes such a signal, being in strong contrast with the green or brown tints and retiring habits of the eatable kinds. The subject was brought by me before the Entomological Society (see Proceedings, March 4th, 1867), in order that those members having opportunities for making observations might do so in the following summer; and I also wrote a letter to the *Field* newspaper, begging that some of its readers would co-operate in making observations on what insects were rejected by birds, at the same time fully explaining the great interest and scientific importance of the problem. It is a curious example of how few of the country readers of that paper are at all interested in questions of simple natural history, that I only obtained one answer from a gentleman in Cumberland, who gave me some interesting observations on the general dislike and abhorrence of all birds to the "Gooseberry Caterpillar," probably that of the Magpie-moth (Abraxas grossulariata). Neither young pheasants, partridges, nor wild-ducks could be induced to eat it, sparrows and finches never touched it, and all birds to whom he offered it rejected it with evident dread and abhorrence. It will be seen that these observations are confirmed by those of two members of the Entomological Society to whom we are indebted for more detailed information.

In March, 1869, Mr. J. Jenner Weir communicated a valuable series of observations made during many years, but more especially in the two preceding summers, in his aviary, containing the following birds of more or less insectivorous habits :—Robin, Yellow - Hammer, Reed-bunting, Bullfinch, Chaffinch, Crossbill, Thrush, Tree-Pipit, Siskin, and Redpoll. He found that hairy caterpillars were uniformly rejected; five distinct species were quite unnoticed by all his birds, and were allowed to crawl about the aviary for days with impunity. The spiny caterpillars of the Tortoiseshell and Peacock but-

terflies were equally rejected ; but in both these cases
Mr. Weir thinks it is the taste, not the hairs or spines,
that are disagreeable, because some very young caterpillars of a hairy species were rejected although no hairs
were developed, and the smooth pupæ of the above-
named butterflies were refused as persistently as the
spined larvæ.   In these cases, then, both hairs and
spines would seem to be mere signs of uneatableness.

His next experiments were with those smooth gaily-
coloured caterpillars which never conceal themselves,
but on the contrary appear to court observation.   Such
are those of the Magpie moth (Abraxas grossulariata),
whose caterpillar is conspicuously white and black
spotted — the Diloba cœruleocephala, whose larvæ is
pale yellow with a broad blue or green lateral band—
the Cucullia verbasci, whose larvæ is greenish white
with yellow bands and black spots, and Anthrocera
filipendulæ (the six spot Burnet moth), whose caterpillar is yellow with black spots.   These were given
to the birds at various times, sometimes mixed with
other kinds of larvæ which were greedily eaten, but
they were in every case rejected apparently unnoticed,
and were left to crawl about till they died.

The next set of observations were on the dull-
coloured and protected larvæ, and the results of numerous experiments are thus summarised by Mr.
Weir.   " All caterpillars whose habits are nocturnal,
which are dull coloured, with fleshy bodies and
smooth skins, are eaten with the greatest avidity.
Every species of green caterpillar is also much re-

lished. All Geometræ, whose larvæ resemble twigs as they stand out from the plant on their anal prolegs, are invariably eaten."

At the same meeting Mr. A. G. Butler, of the British Museum, communicated the results of his observations with lizards, frogs, and spiders, which strikingly corroborate those of Mr. Weir. Three green lizards (Lacerta viridis) which he kept for several years, were very voracious, eating all kinds of food, from a lemon cheesecake to a spider, and devouring flies, caterpillars, and humble bees; yet there were some caterpillars and moths which they would seize only to drop immediately. Among these the principal were the caterpillar of the Magpie moth (Abraxas grossulariata) and the perfect six spot Burnet moth (Anthrocera filipendulæ). These would be first seized but invariably dropped in disgust, and afterwards left unmolested. Subsequently frogs were kept and fed with caterpillars from the garden, but two of these— that of the before-mentioned Magpie moth, and that of the V. moth (Halia wavaria), which is green with conspicuous white or yellow stripes and black spots— were constantly rejected. When these species were first offered, the frogs sprang at them eagerly and licked them into their mouths; no sooner, however, had they done so than they seemed to be aware of the mistake that they had made, and sat with gaping mouths, rolling their tongues about until they had got quit of the nauseous morsels.

With spiders the same thing occurred. These two

caterpillars were repeatedly put into the webs both of
the geometrical and hunting spiders (Epeira diadema
and Lycosa sp.), but in the former case they were
cut out and allowed to drop ; in the latter, after dis-
appearing in the jaws of their captor down his dark
silken funnel, they invariably reappeared, either from
below or else taking long strides up the funnel again.
Mr. Butler has observed lizards fight with and finally
devour humble bees, and a frog sitting on a bed of
stone-crop leap up and catch the bees which flew over
his head, and swallow them, in utter disregard of
their stings.   It is evident, therefore, that the possession of a disagreeable taste or odour is a more effectual protection to certain conspicuous caterpillars and
moths, than would be even the possession of a sting.

The observations of these two gentlemen supply
a very remarkable confirmation of the hypothetical
solution of the difficulty which I had given two years
before.   And as it is generally acknowledged that
the best test of the truth and completeness of a
theory is the power which it gives us of prevision,
we may I think fairly claim this as a case in which
the power of prevision has been successfully exerted,
and therefore as furnishing a very powerful argument in favour of the truth of the theory of Natural
Selection.

Summary.

I have now completed a brief, and necessarily very
imperfect, survey of the various ways in which the

external form and colouring of animals is adapted to be useful to them, either by concealing them from their enemies or from the creatures they prey upon. It has, I hope, been shown that the subject is one of much interest, both as regard a true comprehension of the place each animal fills in the economy of nature, and the means by which it is enabled to maintain that place; and also as teaching us how important a part is played by the minutest details in the structure of animals, and how complicated and delicate is the equilibrium of the organic world.

My exposition of the subject having been necessarily somewhat lengthy and full of details, it will be as well to recapitulate its main points.

There is a general harmony in nature between the colours of an animal and those of its habitation. Arctic animals are white, desert animals are sand-coloured; dwellers among leaves and grass are green; nocturnal animals are dusky. These colours are not universal, but are very general, and are seldom reversed. Going on a little further, we find birds, reptiles, and insects, so tinted and mottled as exactly to match the rock, or bark, or leaf, or flower, they are accustomed to rest upon, — and thereby effectually concealed. Another step in advance, and we have insects which are formed as well as coloured so as exactly to resemble particular leaves, or sticks, or mossy twigs, or flowers; and in these cases very peculiar habits and instincts come into play to aid in the deception and render the concealment more complete. We now enter upon a new phase of the phenomena, and come to creatures whose colours neither conceal them nor make them like vegetable or mineral substances; on the contrary, they are conspicuous enough, but they completely resemble some other creature of a quite different group, while they differ much in outward appearance from those with which all essential parts of their organization show them to be really closely allied. They appear like actors or masqueraders dressed up and painted for amusement, or like swindlers endeavouring to pass themselves off for well-known and respectable members of society. What is the meaning of this strange travestie? Does Nature descend to imposture or masquerade? We answer, she does not. Her principles are too severe. There is a use in every detail of her handiwork. The resemblance of one animal to another is of exactly the same essential nature as the resemblance to a leaf, or to bark, or to desert sand, and answers exactly the same purpose. In the one case the enemy will not attack the leaf or the bark, and so the disguise is a safeguard; in the other case it is found that for various reasons the creature resembled is passed over, and not attacked by the usual enemies of its order, and thus the creature that resembles it has an equally effectual safeguard. We are plainly shown that the disguise is of the same nature in the two cases, by the occurrence in the same group of one species resembling a vegetable substance, while another resembles a living animal of another group; and we know that the creatures re-sembled, possess an immunity from attack, by their being always very abundant, by their being conspi-cuous and not concealing themselves, and by their having generally no visible means of escape from their enemies; while, at the same time, the particular quality that makes them disliked is often very clear, such as a nasty taste or an indigestible hardness. Further examination reveals the fact that, in several cases of both kinds of disguise, it is the female only that is thus disguised; and as it can be shown that the female needs protection much more than the male, and that her preservation for a much longer period is absolutely necessary for the continuance of the race, we have an additional indication that the resemblance is in all cases subservient to a great purpose—the preservation of the species.

In endeavouring to explain these phenomena as having been brought about by variation and natural selection, we start with the fact that white varieties frequently occur, and when protected from enemies show no incapacity for continued existence and in-crease. We know, further, that varieties of many other tints occasionally occur; and as "the survival of the fittest" must inevitably weed out those whose colours are prejudicial and preserve those whose colours are a safeguard, we require no other mode of accounting for the protective tints of arctic and desert animals. But this being granted, there is such a perfectly continuous and graduated series of examples of every kind of protective imitation, up to the most wonderful cases of what is termed "mimicry," that we càn find no place at which to draw the line, and say,—so far variation and natural selection will account for the phenomena, but for all the rest we require a more potent cause. The counter theories that have been proposed, that of the " special creation " of each imitative form, that of the action of "similar conditions of existence " for some of the cases, and of the laws of " hereditary descent and the reversion to ancestral forms " for others,—have all been shown to be beset with difficulties, and the two latter to be directly contradicted by some of the most constant and most remarkable of the facts to be accounted for.

General deductions as to Colour in Nature.

The important part that " protective resemblance " has played in determining the colours and markings of many groups of animals, will enable us to under-stand the meaning of one of the most striking facts in nature, the uniformity in the colours of the vege-table as compared with the wonderful diversity· of the animal world. There appears no good reason why trees and shrubs should not have been adorned with as many varied hues and as strikingly designed pat-terns as birds and butterflies, since the gay colours of flowers show that there is no incapacity in vege-table tissues to exhibit them. But even flowers them-selves present us with none of those wonderful designs, those complicated arrangements of stripes and dots and patches of colour, that harmonious blending of hues in lines and bands and shaded spots, which are so general a feature in insects. It is the opinion of Mr. Darwin that we owe much of the beauty of flowers to the necessity of attracting insects to aid in their fertilisation, and that much of the development of colour in the animal world is due to " sexual selection," colour being universally attractive, and thus leading to its propagation and increase ; but while fully admitting this, it will be evident from the facts and arguments here brought forward, that very much of the *variety* both of colour and markings among animals is due to the supreme importance of concealment, and thus the various tints of minerals and vegetables have been directly reproduced in the animal kingdom, and again and again modified as more special protection became necessary. We shall thus have two causes for the development of colour in the animal world, and shall be better enabled to understand how, by their combined and separate action, the immense variety we now behold has been produced. Both causes, however, will come under the general law of " Utility," the advocacy of which, in its broadest sense, we owe almost entirely to Mr. Darwin. A more accurate knowledge of the varied phenomena connected with this subject may not improbably give us some information both as to the senses and the mental faculties of the lower animals. For it is evident that if colours which please us also attract them, and if the various disguises which have been here enumerated are equally deceptive to them as to ourselves, then both their powers of vision and their faculties of perception and emotion, must be essentially of the same nature as our own—a fact of high philosophical importance in the study of our own nature and our true relations to the lower animals.

Conclusion.

Although such a variety of interesting facts have been already accumulated, the subject we have been discussing is one of which comparatively little is really known. The natural history of the tropics has never yet been studied on the spot with a full appreciation of " what to observe " in this matter. The varied ways in which the colouring and form of animals serve for their protection, their strange disguises as vegetable or mineral substances, their wonderful mimicry of other beings, offer an almost unworked and inexhaustible field of discovery for the zoologist, and will assuredly throw much light on the laws and conditions which have resulted in the wonderful variety of colour, shade, and marking which constitutes one of the most pleasing characteristics of the animal world, but the immediate causes of which it has hitherto been most difficult to explain.

If I have succeeded in showing that in this wide and picturesque domain of nature, results which have hitherto been supposed to depend either upon those incalculable combinations of laws which we term chance or upon the direct volition of the Creator, are really due to the action of comparatively well-known and simple causes, I shall have attained my present purpose, which has been to extend the interest so generally felt in the more striking facts of natural history to a large class of curious but much neglected details; and to further, in however slight a degree, our knowledge of the subjection of the phenomena of life to the " Reign of Law."

IV.

THE MALAYAN PAPILIONIDÆ OR SWALLOW-TAILED BUTTERFLIES, AS ILLUSTRATIVE OF THE THEORY OF NATURAL SELECTION.

Special Value of the Diurnal Lepidoptera for enquiries of this nature.

WHEN the naturalist studies the habits, the structure, or the affinities of animals, it matters little to which group he especially devotes himself; all alike offer him endless materials for observation and research. But, for the purpose of investigating the phenomena of geographical distribution and of local, sexual, or general variation, the several groups differ greatly in their value and importance. Some have too limited a range, others are not sufficiently varied in specific forms, while, what is of most importance, many groups have not received that amount of attention over the whole region they inhabit, which could furnish materials sufficiently approaching to completeness to enable us to arrive at any accurate conclusions as to the phenomena they present as a whole. It is in those groups which are, and have long been, favourites with collectors, that the student of distribution and variation will find his materials the most satisfactory, from their comparative completeness.

Pre-eminent among such groups are the diurnal Lepidoptera or Butterflies, whose extreme beauty and endless diversity have led to their having been assiduously collected in all parts of the world, and to the numerous species and varieties having been figured in a series of magnificent works, from those of Cramer, the contemporary of Linnæus, down to the inimitable productions of our own Hewitson.* But, besides their abundance, their universal distribution, and the great attention that has been paid to them, these insects have other qualities that especially adapt them to elucidate the branches of inquiry already alluded to. These are, the immense development and peculiar structure of the wings, which not only vary in form more than those of any other insects, but offer on both surfaces an endless variety of pattern, colouring, and texture. The scales, with which they are more or less completely covered, imitate the rich hues and delicate surfaces of satin or of velvet, glitter with metallic lustre, or glow with the changeable tints of the opal. This delicately painted surface acts as a register of the minutest differences of organization—a shade of colour, an additional streak or spot, a slight modification of outline continually recurring with the greatest regularity and fixity, while the body and all its other

---

* W. C. Hewitson, Esq., of Oatlands, Walton-on-Thames, author of "Exotic Butterflies" and several other works, illustrated by exquisite coloured figures drawn by himself; and owner of the finest collection of Butterflies in the world.

members exhibit no appreciable change.   The wings of Butterflies, as Mr. Bates has well put it, "serve as a tablet on which Nature writes the story of the modifications of species;" they enable us to perceive changes that would otherwise be uncertain and difficult of observation, and exhibit to us on an enlarged scale the effects of the climatal and other physical conditions which influence more or less profoundly the organization of every living thing.

A proof that this greater sensibility to modifying causes is not imaginary may, I think, be drawn from the consideration, that while the Lepidoptera as a whole are of all insects the least essentially varied in form, structure, or habits, yet in the number of their specific forms they are not much inferior to those orders which range over a much wider field of nature, and exhibit more deeply seated structural modifications.   The Lepidoptera are all vegetable-feeders in their larva-state, and suckers of juices or other liquids in their perfect form.   In their most widely separated groups they differ but little from a common type, and offer comparatively unimportant modifications of structure or of habits.   The Coleoptera, the Diptera, or the Hymenoptera, on the other hand, present far greater and more essential variations.   In either of these orders we have both vegetable and animal-feeders, aquatic, and terrestrial, and parasitic groups. Whole families are devoted to special departments in the economy of nature.   Seeds, fruits, bones, carcases, excrement, bark, have each their special and dependent insect tribes from among them; whereas the Lepidoptera are, with but few exceptions, confined to the one function of devouring the foliage of living vegetation. We might therefore anticipate that their species - population would be only equal to that of sections of the other orders having a similar uniform mode of existence; and the fact that their numbers are at all comparable with those of entire orders, so much more varied in organization and habits, is, I think, a proof that they are in general highly susceptible of specific modification.

Question of the rank of the Papilionidæ.

The Papilionidæ are a family of diurnal Lepidoptera which have hitherto, by almost universal consent, held the first rank in the order; and though this position has recently been denied them, I cannot altogether acquiesce in the reasoning by which it has been proposed to degrade them to a lower rank. In Mr. Bates's most excellent paper on the Heliconidæ, (published in the Transactions of the Linnæan Society, vol. xxiii., p. 495) he claims for that family the highest position, chiefly because of the imperfect structure of the fore legs, which is there carried to an extreme degree of abortion, and thus removes them further than any other family from the Hesperidæ and Heterocera, which all have perfect legs. Now it is a question whether any amount of difference which is exhibited merely in the imperfection or abortion of certain organs, can establish in the

group exhibiting it a claim to a high grade of organization ; still less can this be allowed when another
group along with perfection of structure in the same
organs, exhibits modifications peculiar to it, together
with the possession of an organ which in the re-
mainder of the order is altogether wanting. This is,
however, the position of the Papilionidæ. The perfect insects possess two characters quite peculiar to
them. Mr. Edward Doubleday, in his " Genera of
Diurnal Lepidoptera," says, " The Papilionidæ may
be known by the apparently four-branched median
nervule and the spur on the anterior tibiæ, characters found in no other family." The four-branched
median nervule is a character so constant, so peculiar, and so well marked, as to enable a person to
tell, at a glance at the wings only of a butterfly,
whether it does or does not belong to this family ;
and I am not aware that any other group of butterflies, at all comparable to this in extent and modifications of form, possesses a character in its neuration
to which the same degree of certainty can be attached.
The spur on the anterior tibiæ is also found in some
of the Hesperidæ, and is therefore supposed to show a
direct affinity between the two groups : but I do not
imagine it can counterbalance the differences in neuration and in every other part of their organization.
The most characteristic feature of the Papilionidæ,
however, and that on which I think insufficient
stress has been laid, is undoubtedly the peculiar
structure of the larvæ. These all possess an extra-

ordinary organ situated on the neck, the well-known Y-shaped tentacle, which is entirely concealed in a state of repose, but which is capable of being suddenly thrown out by the insect when alarmed. When we consider this singular apparatus, which in some species is nearly half an inch long, the arrangement of muscles for its protrusion and retraction, its perfect concealment during repose, its blood-red colour, and the suddenness with which it can be thrown out, we must, I think, be led to the conclusion that it serves as a protection to the larva, by startling and frightening away some enemy when about to seize it, and is thus one of the causes which has led to the wide extension and maintained the permanence of this now dominant group. Those who believe that such peculiar structures can only have arisen by very minute successive variations, each one advantageous to its possessor, must see, in the possession of such an organ by one group, and its complete absence in every other, a proof of a very ancient origin and of very long-continued modification. And such a positive structural addition to the organization of the family, subserving an important function, seems to me alone sufficient to warrant us in considering the Papilionidæ as the most highly developed portion of the whole order, and thus in retaining it in the position which the size, strength, beauty, and general structure of the perfect insects have been generally thought to deserve.

In Mr. Trimen's paper on "Mimetic Analogies

among African Butterflies," in the Transactions of the
Linnæan Society, for 1868, he has argued strongly
in favour of Mr. Bates' views as to the higher position of the Danaidæ and the lower grade of the
Papilionidæ, and has adduced, among other facts, the
undoubted resemblance of the pupa of Parnassius, a
genus of Papilionidæ, to that of some Hesperidæ and
moths. I admit, therefore, that he has proved the
Papilionidæ to have retained several characters of
the nocturnal Lepidoptera which the Danaidæ have
lost, but I deny that they are therefore to be considered lower in the scale of organization. Other
characters may be pointed out which indicate that
they are farther removed from the moths even than
the Danaidæ. The club of the antennæ is the most
prominent and most constant feature by which butterflies may be distinguished from moths, and of
all butterflies the Papilionidæ have the most beautiful and most perfectly developed clubbed antennæ.
Again, butterflies and moths are broadly characterised by their diurnal and nocturnal habits respectively,
and the Papilionidæ, with their close allies the Pieridæ, are the most pre-eminently diurnal of butterflies,
most of them lovers of sunshine, and not presenting
a single crepuscular species. The great group of the
Nymphalidæ, on the other hand (in which Mr. Bates
includes the Danaidæ and Heliconidæ as sub-families), contains an entire sub-family (Brassolidæ) and
a number of genera, such as Thaumantis, Zeuxidia,
Pavonia, &c., of crepuscular habits, while a large

proportion of the Satyridæ and many of the Danaidæ are shade-loving butterflies. This question, of
what is to be considered the highest type of any
group of organisms, is one of such general interest to
naturalists that it will be well to consider it a little
further, by a comparison of the Lepidoptera with some
groups of the higher animals.

Mr. Trimen's argument, that the lepidopterous type,
like that of birds, being pre-eminently aërial, " therefore a diminution of the ambulatory organs, instead
of being a sign of inferiority, may very possibly indicate a higher, because a more thoroughly aërial
form," is certainly unsound, for it would imply that
the most aërial of birds (the swift and the frigate-
birds, for example) are the highest in the scale of
bird-organization, and the more so on account of their
feet being very ill adapted for walking. But no or-
nithologist has ever so classed them, and the claim to
the highest rank among birds is only disputed between three groups, all very far removed from these.
They are — 1st. The Falcons, on account of their
general perfection, their rapid flight, their piercing
vision, their perfect feet armed with retractile claws,
the beauty of their forms, and the ease and rapidity of
their motions; 2nd. The Parrots, whose feet, though
ill-fitted for walking, are perfect as prehensile organs,
and which possess large brains with great intelligence,
though but moderate powers of flight; and, 3rd. The
Thrushes or Crows, as typical of the perching birds,
on account of the well-balanced development of their

whole structure, in which no organ or function has attained an undue prominence.

Turning now to the Mammalia, it might be argued that as they are pre-eminently the terrestrial type of vertebrates, to walk and run well is essential to the typical perfection of the group; but this would give the superiority to the horse, the deer, or the hunting leopard, instead of to the Quadrumana. We seem here to have quite a case in point, for one group of Quadrumana, the Lemurs, is undoubtedly nearer to the low Insectivora and Marsupials than the Carnivora or the Ungulata, as shown among other characters by the Opossums possessing a hand with perfect opposable thumb, closely resembling that of some of the Lemurs; and by the curious Galeopithecus, which is sometimes classed as a Lemur, and sometimes with the Insectivora. Again, the implacental mammals, including the Ornithodelphia and the Marsupials, are admitted to be lower than the placental series. But one of the distinguishing characters of the Marsupials is that the young are born blind and exceedingly imperfect, and it might therefore be argued that those orders in which the young are born most perfect are the highest, because farthest from the low Marsupial type. This would make the Ruminants and Ungulata higher than the Quadrumana or the Carnivora. But the Mammalia offer a still more remarkable illustration of the fallacy of this mode of reasoning, for if there is one character more than another which is essential and distinctive of the class, it is that from which it derives its name, the possession of mammary glands and the power of suckling the young. What more reasonable, apparently, than to argue that the group in which this important function is most developed, that in which the young are most dependent upon it, and for the longest period, must be the highest in the Mammalian scale of organization? Yet this group is the Marsupial, in which the young commence suckling in a fœtal condition, and continue to do so till they are fully developed, and are therefore for a long time absolutely dependent on this mode of nourishment.

These examples, I think, demonstrate that we cannot settle the rank of a group by a consideration of the degree in which certain characters resemble or differ from those in what is admitted to be a lower group; and they also show that the highest group of a class may be more closely connected to one of the lowest, than some other groups which have developed laterally and diverged farther from the parent type, but which yet, owing to want of balance or too great specialization in their structure, have never reached a high grade of organization. The Quadrumana afford a very valuable illustration, because, owing to their undoubted affinity with man, we feel certain that they are really higher than any other order of Mammalia, while at the same time they are more distinctly allied to the lowest groups than many others. The case of the Papilionidæ seems to me so exactly parallel to this, that, while I admit all the proofs of affinity with the undoubtedly lower groups of Hesperidæ and moths, I yet maintain that, owing to the complete and even development of every part of their organization, these insects best represent the highest perfection to which the butterfly type has attained, and deserve to be placed at its head in every system of classification.

Distribution of the Papilionidæ.

The Papilionidæ are pretty widely distributed over the earth, but are especially abundant in the tropics, where they attain their maximum of size and beauty, and the greatest variety of form and colouring. South America, North India, and the Malay Islands are the regions where these fine insects occur in the greatest profusion, and where they actually become a not unimportant feature in the scenery. In the Malay Islands in particular, the giant Ornithopteræ may be frequently seen about the borders of the cultivated and forest districts, their large size, stately flight, and gorgeous colouring rendering them even more conspicuous than the generality of birds. In the shady suburbs of the town of Malacca two large and handsome Papilios (Memnon and Nephelus) are not uncommon, flapping with irregular flight along the roadways, or, in the early morning, expanding their wings to the invigorating rays of the sun. In Amboyna and other towns of the Moluccas, the magnificent Deiphobus and Severus, and occasionally even the azure-winged Ulysses, frequent similar situations, fluttering about the orange-trees and flower-beds, or sometimes even straying into the narrow bazaars or covered markets of the city. In Java the golden-dusted Arjuna may often be seen at damp places on the roadside in the mountain districts, in company with Sarpedon, Bathycles, and Agamemnon, and less frequently the beautiful swallow-tailed Antiphates. In the more luxuriant parts of these islands one can hardly take a morning's walk in the neighbourhood of a town or village without seeing three or four species of Papilio, and often twice that number. No less than 130 species of the family are now known to inhabit the Archipelago, and of these ninety-six were collected by myself. Thirty species are found in Borneo, being the largest number in any one island, twenty-three species having been obtained by myself in the vicinity of Sarawak; Java has twenty-eight species; Celebes twenty-four, and the Peninsula of Malacca, twenty-six species. Further east the numbers decrease; Batchian producing seventeen, and New Guinea only fifteen, though this number is certainly too small, owing to our present imperfect knowledge of that great island.

Definition of the word Species.

In estimating these numbers I have had the usual difficulty to encounter, of determining what to consider species and what varieties. The Malayan region, consisting of a large number of islands of generally great antiquity, possesses, compared to its actual area, a great number of distinct forms, often indeed distinguished by very slight characters, but in most cases so constant in large series of specimens, and so easily separable from each other, that I know not on what principle we can refuse to give them the name and rank of species. One of the best and most orthodox definitions is that of Pritchard, the great ethnologist, who says, that " *separate origin and distinctness of race, evinced by a constant transmission of some characteristic peculiarity of organization,*" constitutes a species. Now leaving out the question of " origin," which we cannot determine, and taking only the proof of separate origin, " *the constant transmission of some characteristic peculiarity of organization,*" we have a definition which will compel us to neglect altogether the *amount* of difference between any two forms, and to consider only whether the differences that present themselves are *permanent.* The rule, therefore, I have endeavoured to adopt is, that when the difference between two forms inhabiting separate areas seems quite constant, when it can be defined in words, and when it is not confined to a single peculiarity only, I have considered such forms to be species. When, however, the individuals of each locality vary among themselves, so as to cause the distinctions between the two forms to become inconsiderable and indefinite, or where the differences, though constant, are confined to one particular only, such as size, tint, or a single point of difference in marking or in outline, I class one of the forms as a variety of the other.

I find as a general rule that the constancy of species is in an inverse ratio to their range. Those which are confined to one or two islands are generally very constant. When they extend to many islands, considerable variability appears; and when they have an extensive range over a large part of the Archipelago, the amount of unstable variation is very large. These facts are explicable on Mr. Darwin's principles. When a species exists over a wide area, it must have had, and probably still possesses, great powers of dispersion. Under the different conditions of existence in various portions of its area, different variations from the type would be selected, and, were they completely isolated, would soon become distinctly modified forms; but this process is checked by the dispersive powers of the whole species, which leads to the more or less frequent intermixture of the incipient varieties, which thus become irregular and unstable. Where, however, a species has a limited range, it indicates less active powers of dispersion, and the process of modification under changed conditions is less interfered with. The species will therefore exist under one or more permanent forms according as portions of it have been isolated at a more or less remote period.

Laws and Modes of Variation.

What is commonly called variation consists of several distinct phenomena which have been too often confounded. I shall proceed to consider these under the heads of—1st, simple variability; 2nd, polymorphism;

3rd, local forms ; 4th, co-existing varieties ; 5th, races or subspecies ; and 6th, true species.

1. *Simple variability.*—Under this head I include all those cases in which the specific form is to some extent unstable. Throughout the whole range of the species, and even in the progeny of individuals, there occur continual and uncertain differences of form, analogous to that variability which is so characteristic of domestic breeds. It is impossible usefully to define any of these forms, because there are indefinite gradations to each other form. Species which possess these characteristics have always a wide range, and are more frequently the inhabitants of continents than of islands, though such cases are always exceptional, it being far more common for specific forms to be fixed within very narrow limits of variation. The only good example of this kind of variability which occurs among the Malayan Papilionidæ is in Papilio Severus, a species inhabiting all the islands of the Moluccas and New Guinea, and exhibiting in each of them a greater amount of individual difference than often serves to distinguish well-marked species. Almost equally remarkable are the variations exhibited in most of the species of Ornithoptera, which I have found in some cases to extend even to the form of the wing and the arrangement of the nervures. Closely allied, however, to these variable species are others which, though differing slightly from them, are constant and confined to limited areas. After satisfying oneself, by the examination of numerous specimens captured in their native countries, that the

one set of individuals are variable and the others are not, it becomes evident that by classing all alike as varieties of one species we shall be obscuring an important fact in nature; and that the only way to exhibit that fact in its true light is to treat the invariable local form as a distinct species, even though it does not offer better distinguishing characters than do the extreme forms of the variable species. Cases of this kind are the Ornithoptera Priamus, which is confined to the islands of Ceram and Amboyna, and is very constant in both sexes, while the allied species inhabiting New Guinea and the Papuan Islands is exceedingly variable; and in the island of Celebes is a species closely allied to the variable P. Severus, but which, being exceedingly constant, I have described as a distinct species under the name of Papilio Pertinax.

2. *Polymorphism or dimorphism.*—By this term I understand the co-existence in the same locality of two or more distinct forms, not connected by intermediate gradations, and all of which are occasionally produced from common parents. These distinct forms generally occur in the female sex only, and their offspring, instead of being hybrids, or like the two parents, appear to reproduce all the distinct forms in varying proportions. I believe it will be found that a considerable number of what have been classed as *varieties* are really cases of polymorphism. Albinoism and melanism are of this character, as well as most of those cases in which well-marked varieties occur in company with the parent species, but without any intermediate forms. If

L

these distinct forms breed independently, and are never reproduced from a common parent, they must be considered as separate species, contact without intermixture being a good test of specific difference. On the other hand, intercrossing without producing an intermediate race is a test of dimorphism. I consider, therefore, that under any circumstances. the term " variety " is wrongly applied to such cases.

The Malayan Papilionidæ exhibit some very curious instances of polymorphism, some of which have been recorded as varieties, others as distinct species; and they all occur in the female sex. Papilio Memnon is one of the most striking, as it exhibits the mixture of simple variability, local and polymorphic forms, all hitherto classed under the common title of varieties. The polymorphism is strikingly exhibited by the females, one set of which resemble the males in form, with a variable paler colouring ; the others have a large spatulate tail to the hinder wings and a distinct style of colouring, which causes them closely to resemble P. Coon, a species having the two sexes alike and inhabiting the same countries, but with which they have no direct affinity. The tailless females exhibit simple variability, scarcely two being found exactly alike even in the same locality. The males of the island of Borneo exhibit constant differences of the under surface, and may therefore be distinguished as a local form, while the continental specimens, as a whole, offer such large and constant differences from those of the islands, that I am inclined to separate them as a distinct species, to which the name P. Androgeus (Cramer) may be applied. We have here, therefore, distinct species, local forms, polymorphism, and simple variability, which seem to me to be distinct phenomena, but which have been hitherto all classed together as varieties. I may mention that the fact of these distinct forms being one species is doubly proved. The males, the tailed and tailless females, have all been bred from a single group of the larvæ, by Messrs. Payen and Bocarmé, in Java, and I myself captured, in Sumatra, a male P. Memnon, and a tailed female P. Achates, under circumstances which led me to class them as the same species.

Papilio Pammon offers a somewhat similar case. The female was described by Linnæus as P. Polytes, and was considered to be a distinct species till Westermann bred the two from the same larvæ (see Boisduval, "Species Général des Lépidoptères," p. 272). They were therefore classed as sexes of one species by Mr. Edward Doubleday, in his "Genera of Diurnal Lepidoptera," in 1846. Later, female specimens were received from India closely resembling the male insect, and this was held to overthrow the authority of M. Westermann's observation, and to re-establish P. Polytes as a distinct species; and as such it accordingly appears in the British Museum List of Papilionidæ in 1856, and in the Catalogue of the East India Museum in 1857. This discrepancy is explained by the fact of P. Pammon having two females, one closely resembling the male, while the other is totally different from it. A long familiarity with this insect (which

replaced by local forms or by closely allied species, occurs
in every island of the Archipelago) has convinced me
of the correctness of this statement; for in every place
where a male allied to P. Pammon is found, a female
resembling P. Polytes also occurs, and sometimes,
though less frequently than on the continent, another
female closely resembling the male : while not only has
no male specimen of P. Polytes yet been discovered,
but the female (Polytes) has never yet been found in
localities to which the male (Pammon) does not extend.
In this case, as in the last, distinct species, local forms,
and dimorphic specimens, have been confounded under
the common appellation of varieties.

But, besides the true P. Polytes, there are several
allied forms of females to be considered, namely, P.
Theseus (Cramer), P. Melanides (De Haan), P. Elyros
(G. R. Gray), and P. Romulus (Linnæus). The dark
female figured by Cramer as P. Theseus seems to be
the common and perhaps the only form in Sumatra,
whereas in Java, Borneo, and Timor, along with males
quite identical with those of Sumatra, occur females
of the Polytes form, although a single specimen of
the true P. Theseus taken at Lombock would seem to
show that the two forms do occur together. In the
allied species found in the Philippine Islands (P. Alphenor, Cramer = P. Ledebouria, Eschscholtz, the
female of which is P. Elyros, G. R. Gray,) forms
corresponding to these extremes occur, along with a
number of intermediate varieties, as shown by a fine
series in the British Museum. We have here an

indication of how dimorphism may be produced; for let the extreme Philippine forms be better suited to their conditions of existence than the intermediate connecting links, and the latter will gradually die out, leaving two distinct forms of the same insect, each adapted to some special conditions. As these conditions are sure to vary in different districts, it will often happen, as in Sumatra and Java, that the one form will predominate in the one island, the other in the adjacent one. In the island of Borneo there seems to be a third form; for P. Melanides (De Haan) evidently belongs to this group, and has all the chief characteristics of P. Theseus, with a modified colouration of the hind wings. I now come to an insect which, if I am correct, offers one of the most interesting cases of variation yet adduced. Papilio Romulus, a butterfly found over a large part of India and Ceylon, and not uncommon in collections, has always been considered a true and independent species, and no suspicions have been expressed regarding it. But a male of this form does not, I believe, exist. I have examined the fine series in the British Museum, in the East India Company's Museum, in the Hope Museum at Oxford, in Mr. Hewitson's and several other private collections, and can find nothing but females; and for this common butterfly no male partner can be found except the equally common P. Pammon, a species already provided with two wives, and yet to whom we shall be forced, I believe, to assign a third. On carefully examining P. Romulus,

I find that in all essential characters—the form and texture of the wings, the length of the antennæ, the spotting of the head and thorax, and even the peculiar tints and shades with which it is ornamented—it corresponds exactly with the other females of the Pammon group ; and though, from the peculiar marking of the fore wings, it has at first sight a very different aspect, yet a closer examination shows that every one of its markings could be produced by slight and almost imperceptible modifications of the various allied forms.   I fully believe, therefore, that I shall be correct in placing P. Romulus as a third Indian form of the female P. Pammon, corresponding to P. Melanides, the third form of the Malayan P. Theseus. I may mention here that the females of this group have a superficial resemblance to the Polydorus group of Papilios, as shown by P. Theseus having been considered to be the female of P. Antiphus, and by P. Romulus being arranged next to P. Hector.   There is no close affinity between these two groups of Papilio, and I am disposed to believe that we have here a case of mimicry, brought about by the same causes which Mr. Bates has so well explained in his account of the Heliconidæ, and which has led to the singular exuberance of polymorphic forms in this and allied groups of the genus Papilio.   I shall have to devote a section of my essay to the consideration of this subject.

The third example of polymorphism I have to bring forward is Papilio Ormenus, which is closely allied to the well-known P. Erechtheus, of Australia. The most common form of the female also resembles that of P. Erechtheus; but a totally different-looking insect was found by myself in the Aru Islands, and figured by Mr. Hewitson under the name of P. Onesimus, which subsequent observation has convinced me is a second form of the female of P. Ormenus. Comparison of this with Boisduval's description of P. Amanga, a specimen of which from New Guinea is in the Paris Museum, shows the latter to be a closely similar form; and two other specimens were obtained by myself, one in the island of Goram and the other in Waigiou, all evidently local modifications of the same form. In each of these localities males and ordinary females of P. Ormenus were also found. So far there is no evidence that these light-coloured insects are not females of a distinct species, the males of which have not been discovered. But two facts have convinced me this is not the case. At Dorey, in New Guinea, where males and ordinary females closely allied to P. Ormenus occur (but which seem to me worthy of being separated as a distinct species), I found one of these light-coloured females closely followed in her flight by three males, exactly in the same manner as occurs (and, I believe, occurs only) with the sexes of the same species. After watching them a considerable time, I captured the whole of them, and became satisfied that I had discovered the true relations of this anomalous form. The next year I had corroborative proof of the correctness of this opinion by the discovery in the island of Batchian of a new species allied to P. Ormenus, all the females of which, either seen or captured by me, were of one form, and much more closely resembling the abnormal light-coloured females of P. Ormenus and P. Pandion than the ordinary specimens of that sex. Every naturalist will, I think, agree that this is strongly confirmative of the supposition that both forms of female are of one species; and when we consider, further, that in four separate islands, in each of which I resided for several months, the two forms of female were obtained and only one form of male ever seen, and that about the same time, M. Montrouzier in Woodlark Island, at the other extremity of New Guinea (where he resided several years, and must have obtained all the large Lepidoptera of the island), obtained females closely resembling mine, which, in despair at finding no appropriate partners for them, he mates with a widely different species—it becomes, I think, sufficiently evident this is another case of polymorphism of the same nature as those already pointed out in P. Pammon and P. Memnon. This species, however, is not only dimorphic, but trimorphic; for, in the island of Waigiou, I obtained a third female quite distinct from either of the others, and in some degree intermediate between the ordinary female and the male. The specimen is particularly interesting to those who believe, with Mr. Darwin, that extreme difference of the sexes has been gradually produced by what he terms sexual selection, since it may be supposed to exhibit one of the intermediate steps in that process, which has been accidentally preserved in company with its more favoured rivals, though its extreme rarity (only one specimen having been seen to many hundreds of the other form) would indicate that it may soon become extinct.

The only other case of polymorphism in the genus Papilio, at all equal in interest to those I have now brought forward, occurs in America; and we have, fortunately, accurate information about it. Papilio Turnus is common over almost the whole of temperate North America; and the female resembles the male very closely. A totally different-looking insect both in form and colour, Papilio Glaucus, inhabits the same region; and though, down to the time when Boisduval published his " Species Général," no connexion was supposed to exist between the two species, it is now well ascertained that P. Glaucus is a second female form of P. Turnus. In the "Proceedings of the Entomological Society of Philadelphia," Jan., 1863, Mr. Walsh gives a very interesting account of the distribution of this species. He tells us that in the New England States and in New York all the females are yellow, while in Illinois and further south all are black; in the intermediate region both black and yellow females occur in varying proportions. Lat. 37° is approximately the southern limit of the yellow form, and 42° the northern limit of the black form; and, to render the proof complete, both black and yellow insects have been bred from a single batch of eggs. He further states that, out of thousands of specimens, he has never seen or heard of intermediate varieties between these forms. In this interesting example we see the effects of latitude in determining the proportions in which the individuals of each form should exist. The conditions are *here* favourable to the one form, *there* to the other; but we are by no means to suppose that these conditions consist in climate alone. It is highly probable that the existence of enemies, and of competing forms of life, may be the main determining influences; and it is much to be wished that such a competent observer as Mr. Walsh would endeavour to ascertain what are the adverse causes which are most efficient in keeping down the numbers of each of these contrasted forms.

Dimorphism of this kind in the animal kingdom does not seem to have any direct relations to the reproductive powers, as Mr. Darwin has shown to be the case in plants, nor does it appear to be very general. One other case only is known to me in another family of my eastern Lepidoptera, the Pieridæ; and but few occur in the Lepidoptera of other countries. The spring and autumn broods of some European species differ very remarkably; and this must be considered as a phenomenon of an analogous though not of an identical nature, while the Araschnia prorsa, of Central Europe, is a striking example of this alternate or seasonal dimorphism. Among our nocturnal Lepidoptera, I am informed, many analogous cases occur; and as the whole history of many of these has been investigated by breeding successive generations from the egg, it is to be hoped that some of our British Lepidopterists will give us a connected account of all the abnormal phenomena which they present. Among the Coleoptera Mr. Pascoe has pointed out the existence of two forms of the male sex in seven species of the two genera Xenocerus and Mecocerus belonging to the family Anthribidæ, (Proc. Ent. Soc. Lond., 1862); and no less than six European Water-beetles, of the genus Dytiscus, have females of two forms, the most common having the elytra deeply sulcate, the rarer smooth as in the males. The three, and sometimes four or more, forms under which many Hymenopterous insects (especially Ants) occur, must be considered as a related phenomenon, though here each form is specialized to a distinct function in the economy of the species. Among the higher animals, albinoism and melanism may, as I have already stated, be considered as analogous facts; and I met with one case of a bird, a species of Lory (Eos fuscata), clearly existing under two differently coloured forms, since I obtained both sexes of each from a single flock, while no intermediate specimens have yet been found.

The fact of the two sexes of one species differing very considerably is so common, that it attracted but little attention till Mr. Darwin showed how it could in many cases be explained by the principle of

sexual selection.   For instance, in most polygamous
animals the males fight for the possession of the
females, and the victors, always becoming the progenitors of the succeeding generation, impress upon
their male offspring their 'own superior size, strength,
or unusually developed offensive weapons.   It is thus
that we can account for the spurs and the superior
strength and size of the males in Gallinaceous birds,
and also for the large canine tusks in the males of
fruit-eating Apes.   So the superior beauty of plumage
and special adornments of the males of so many birds
can be explained by supposing (what there are many
facts to prove) that the females prefer the most beautiful and perfect-plumaged males, and that thus, slight
accidental variations of form and colour have been
accumulated, till they have produced the wonderful
train of the Peacock and the gorgeous plumage of
the Bird of Paradise.   Both these causes have no
doubt acted partially in insects, so many species
possessing horns and powerful jaws in the male sex
only, and still more frequently the males alone re-
joicing in rich colours or sparkling lustre.   But there
is here another cause which has led to sexual differences, viz., a special adaptation of the sexes to diverse
habits or modes of life.   This is well seen in female
Butterflies (which are generally weaker and of slower
flight), often having colours better adapted to concealment; and in certain South American species (Papilio torquatus) the females, which inhabit the forests,
resemble the Æneas group of Papilios which abound

in similar localities, while the males, which frequent the sunny open river-banks, have a totally different colouration. In these cases, therefore, natural selection seems to have acted independently of sexual selection; and all such cases may be considered as examples of the simplest dimorphism, since the offspring never offer intermediate varieties between the parent forms.

The phenomena of dimorphism and polymorphism may be well illustrated by supposing that a blue-eyed, flaxen-haired Saxon man had two wives, one a black-haired, red-skinned Indian squaw, the other a woolly-headed, sooty-skinned negress—and that instead of the children being mulattoes of brown or dusky tints, mingling the separate characteristics of their parents in varying degrees, all the boys should be pure Saxon boys like their father, while the girls should altogether resemble their mothers. This would be thought a sufficiently wonderful fact; yet the phenomena here brought forward as existing in the insect-world are still more extraordinary; for each mother is capable not only of producing male offspring like the father, and female like herself, but also of producing other females exactly like her fellow-wife, and altogether differing from herself. If an island could be stocked with a colony of human beings having similar physiological idiosyncrasies with Papilio Pammon or Papilio Ormenus, we should see white men living with yellow, red, and black women, and their offspring always reproducing the same types; so that at the end of many generations the men would remain pure white, and the women of the same well-marked races as at the commencement.

The distinctive character therefore of dimorphism is this, that the union of these distinct forms does not produce intermediate varieties, but reproduces the distinct forms unchanged. In simple varieties, on the other hand, as well as when distinct local forms or distinct species are crossed, the offspring never resembles either parent exactly, but is more or less intermediate between them. Dimorphism is thus seen to be a specialized result of variation, by which new physiological phenomena have been developed; the two should therefore, whenever possible, be kept separate.

3. *Local form, or variety.*—This is the first step in the transition from variety to species. It occurs in species of wide range, when groups of individuals have become partially isolated in several points of its area of distribution, in each of which a characteristic form has become more or less completely segregated. Such forms are very common in all parts of the world, and have often been classed by one author as varieties, by another as species. I restrict the term to those cases where the difference of the forms is very slight, or where the segregation is more or less imperfect. The best example in the present group is Papilio Agamemnon, a species which ranges over the greater part of tropical Asia, the whole of the Malay archipelago, and a portion of the Australian and Pacific regions. The modifications are principally of size and form,

and, though slight, are tolerably constant in each locality. The steps, however, are so numerous and gradual that it would be impossible to define many of them, though the extreme forms are sufficiently distinct. Papilio Sarpedon presents somewhat similar but less numerous variations.

4. *Co-existing Variety.*—This is a somewhat doubtful case. It is when a slight but permanent and hereditary modification of form exists in company with the parent or typical form, without presenting those intermediate gradations which would constitute it a case of simple variability. It is evidently only by direct evidence of the two forms breeding separately that this can be distinguished from dimorphism. The difficulty occurs in Papilio Jason, and P. Evemon, which inhabit the same localities, and are almost exactly alike in form, size, and colouration, except that the latter always wants a very conspicuous red spot on the under surface, which is found not only in P. Jason, but in all the allied species. It is only by breeding the two insects that it can be determined whether this is a case of a co-existing variety or of dimorphism. In the former case, however, the difference being constant and so very conspicuous and easily defined, I see not how we could escape considering it as a distinct species. A true case of co-existing forms would, I consider, be produced, if a slight variety had become fixed as a local form, and afterwards been brought into contact with the parent species, with little or no intermixture of the two; and such instances do very probably occur.

5. *Race or subspecies.*—These are local forms completely fixed and isolated; and there is no possible test but individual opinion to determine which of them shall be considered as species and which varieties. If stability of form and " *the constant transmission of some characteristic peculiarity of organization* " is the test of a species (and I can find no other test that is more certain than individual opinion) then every one of these fixed races, confined as they almost always are to distinct and limited areas, must be regarded as a species; and as such I have in most cases treated them. The various modifications of Papilio Ulysses, P. Peranthus, P. Codrus, P. Eurypilus, P. Helenus, &c., are excellent examples; for while some present great and well-marked, others offer slight and inconspicuous differences, yet in all cases these differences seem equally fixed and permanent. If, therefore, we call some of these forms species, and others varieties, we introduce a purely arbitrary distinction, and shall never be able to decide where to draw the line. The races of Papilio Ulysses, for example, vary in amount of modification from the scarcely differing New Guinea form to those of Woodlark Island and New Caledonia, but all seem equally constant; and as most of these had already been named and described as species, I have added the New Guinea form under the name of P. Autolycus. We thus get a little group of Ulyssine Papilios, the whole comprised within a very limited area, each one confined to a separate portion of that area, and, though differing in various amounts, each apparently constant.

Few naturalists will doubt that all these may and probably have been derived from a common stock, and therefore it seems desirable that there should be a unity in our method of treating them; either call them all *varieties* or all *species*. Varieties, however, continually get overlooked; in lists of species they are often altogether unrecorded; and thus we are in danger of neglecting the interesting phenomena of variation and distribution which they present. I think it advisable, therefore, to name all such forms; and those who will not accept them as species may consider them as sub-species or races.

6. *Species.* — Species are merely those strongly marked races or local forms which when in contact do not intermix, and when inhabiting distinct areas are generally believed to have had a separate origin, and to be incapable of producing a fertile hybrid offspring. But as the test of hybridity cannot be applied in one case in ten thousand, and even if it could be applied would prove nothing, since it is founded on an assumption of the very question to be decided—and as the test of separate origin is in every case inapplicable — and as, further, the test of non-intermixture is useless, except in those rare cases where the most closely allied species are found inhabiting the same area, it will be evident that we have no means whatever of distinguishing so-called " true species " from the several modes of variation here pointed out, and into which they so often pass by an insensible gradation. It is quite true that, in

the great majority of cases, what we term " species " are so well marked and definite that there is no difference of opinion about them ; but as the test of a true theory is, that it accounts for, or at the very least is not inconsistent with, the whole of the phenomena and apparent anomalies of the problem to be solved, it is reasonable to ask that those who deny the origin of species by variation and selection should grapple with the facts in detail, and show how the doctrine of the distinct origin and permanence of species will explain and harmonize them. It has been recently asserted by Dr. J. E. Gray (in the Proceedings of the Zoological Society for 1863, page 134), that the difficulty of limiting species is in proportion to our ignorance, and that just as groups or countries are more accurately known and studied in greater letail the limits of species become settled. This statement has, like many other general assertions, its portion of both truth and error. There is no doubt that many uncertain species, founded on few or isolated specimens, have had their true nature determined by the study of a good series of examples : they have been thereby established as species or as varieties ; and the number of times this has occurred is doubtless very great. But there are other, and equally trustworthy cases, in which, not single species, but whole groups have, by the study of a vast accumulation of materials, been proved to have no definite specific limits. A few of these must be adduced. In Dr. Carpenter's " Introduction to the Study of the Foraminifera," he states that " *there is not a single specimen of plant or animal of which the range of variation has been studied by the collocation and comparison of so large a number of specimens as have passed under the review of Messrs. Williamson, Parker, Rupert Jones, and myself, in our studies of the types of this group;* " and the result of this extended comparison of specimens is stated to be, " *The range of variation is so great among the Foraminifera as to include not merely those differential characters which have been usually accounted* SPECIFIC, *but also those upon which the greater part of the* GENERA *of this group have been founded, and even in some instances those of its* ORDERS " (Foraminifera, Preface, x). Yet this same group had been divided by D'Orbigny and other authors into a number of clearly defined *families, genera,* and *species,* which these careful and conscientious researches have shown to have been almost all founded on incomplete knowledge.

Professor DeCandolle has recently given the results of an extensive review of the species of Cupuliferæ. He finds that the best-known species of oaks are those which produce most varieties and subvarieties; that they are often surrounded by provisional species; and, with the fullest materials at his command, two-thirds of the species he considers more or less doubtful. His general conclusion is, that " *in botany the lowest series of groups,* SUBVARIETIES, VARIETIES, *and* RACES *are very badly limited; these can be grouped into* SPECIES *a little less vaguely limited, which again can be formed into sufficiently precise* GENERA." This

general conclusion is entirely objected to by the writer
of the article in the " Natural History Review," who,
however, does not deny its applicability to the particular order under discussion, while this very difference of opinion is another proof that difficulties in
the determination of species do not, any more than
in the higher groups, vanish with increasing materials and more accurate research.

Another striking example of the same kind is seen
in the genera Rubus and Rosa, adduced by Mr.
Darwin himself; for though the amplest materials
exist for a knowledge of these groups, and the most
careful research has been bestowed upon them, yet
the various species have not thereby been accurately
limited and defined so as to satisfy the majority of
botanists.   In Mr. Baker's revision of the British
Roses, just published by the Linnæan Society, the
author includes under the single species Rosa canina,
no less than twenty-eight named *varieties*, distinguished by more or less constant characters and often
confined to special localities ; and to these are referred
about seventy of the *species* of Continental and British
botanists.

Dr. Hooker seems to have found the same thing
in his study of the Arctic flora. For though he has
had much of the accumulated materials of his predecessors to work upon, he continually expresses himself as unable to do more than group the numerous
and apparently fluctuating forms into more or less imperfectly defined species. In his paper on the " Dis-

tribution of Arctic Plants," (Trans. Linn. Soc. xxiii., p. 310) Dr. Hooker says:—"The most able and experienced descriptive botanists vary in their estimate of the value of the 'specific term' to a much greater extent than is generally supposed." . . "I think I may safely affirm that the 'specific term' has three different standard values, all current in descriptive botany, but each more or less confined to one class of observers." . . "This is no question of what is right or wrong as to the real value of the specific term; I believe each is right according to the standard he assumes as the specific."

Lastly, I will adduce Mr. Bates's researches on the Amazons. During eleven years he accumulated vast materials, and carefully studied the variation and distribution of insects. Yet he has shown that many species of Lepidoptera, which before offered no special difficulties, are in reality most intricately combined in a tangled web of affinities, leading by such gradual steps from the slightest and least stable variations to fixed races and well-marked species, that it is very often impossible to draw those sharp dividing-lines which it is supposed that a careful study and full materials will always enable us to do.

These few examples show, I think, that in every department of nature there occur instances of the instability of specific form, which the increase of materials aggravates rather than diminishes. And it must be remembered that the naturalist is rarely likely to err on the side of imputing greater indefiniteness to species than really exists. There is a completeness and satisfaction to the mind in defining and limiting and naming a species, which leads us all to do so whenever we conscientiously can, and which we know has led many collectors to reject vague intermediate forms as destroying the symmetry of their cabinets. We must therefore consider these cases of excessive variation and instability as being thoroughly well established; and to the objection that, after all, these cases are but few compared with those in which species can be limited and defined, and are therefore merely exceptions to a general rule, I reply that a true law embraces all apparent exceptions, and that to the great laws of nature there are no real exceptions—that what appear to be such are equally results of law, and are often (perhaps indeed always) those very results which are most important as revealing the true nature and action of the law. It is for such reasons that naturalists now look upon the study of *varieties* as more important than that of well-fixed species. It is in the former that we see nature still at work, in the very act of producing those wonderful modifications of form, that endless variety of colour, and that complicated harmony of relations, which gratify every sense and give occupation to every faculty of the true lover of nature.

Variation as specially influenced by Locality.

The phenomena of variation as influenced by locality have not hitherto received much attention. Botanists, it is true, are acquainted with the influences of climate, altitude, and other physical conditions, in modifying the forms and external characteristics of plants; but I am not aware that any peculiar influence has been traced to locality, independent of climate. Almost the only case I can find recorded is mentioned in that repertory of natural-history facts, "The Origin of Species," viz. that herbaceous groups have a tendency to become arboreal in islands. In the animal world, I cannot find that any facts have been pointed out as showing the special influence of locality in giving a peculiar *facies* to the several disconnected species that inhabit it. What I have to adduce on this matter will therefore, I hope, possess some interest and novelty.

On examining the closely allied species, local forms, and varieties distributed over the Indian and Malayan regions, I find that larger or smaller districts, or even single islands, give a special character to the majority of their Papilionidæ. For instance: 1. The species of the Indian region (Sumatra, Java, and Borneo) are almost invariably smaller than the allied species inhabiting Celebes and the Moluccas; 2. The species of New Guinea and Australia are also, though in a less degree, smaller than the nearest species or varieties of the Moluccas; 3. In the Moluccas themselves the species of Amboyna are the largest; 4. The species of Celebes equal or even surpass in size those of Amboyna; 5. The species and varieties of Celebes possess a striking character in the form of

the anterior wings, different from that of the allied
species and varieties of all the surrounding islands;
6. Tailed species in India or the Indian region become
tailless as they spread eastward through the archipelago; 7. In Amboyna and Ceram the females of
several species are dull-coloured, while in the adjacent
islands they are more brilliant.

Local variation of Size.—Having preserved the finest
and largest specimens of Butterflies in my own collection, and having always taken for comparison the
largest specimens of the same sex, I believe that the
tables I now give are sufficiently exact. The differences
of expanse of wings are in most cases very great, and
are much more conspicuous in the specimens themselves
than on paper. It will be seen that no less than fourteen Papilionidæ inhabiting Celebes and the Moluccas
are from one-third to one-half greater in extent of wing
than the allied species representing them in Java, Sumatra, and Borneo. Six species inhabiting Amboyna
are larger than the closely allied forms of the northern
Moluccas and New Guinea by about one-sixth. These
include almost every case in which closely allied
species can be compared.

Species of Papilionidæ of the Moluccas and Celebes (large).	Expanse. Inches.	Closely allied species of Java and the Indian region (small).	Expanse. Inches.
Ornithoptera Helena Amboyna)	7·6	O. Pompeus	5·8
		O. Amphrisius	6·0
Papilio Adamantius (Celebes)	5·8		
P. Lorquinianus (Mo-luccas)	4·8	P. Peranthus	3·8

Species of Papilionidæ of the Moluccas and Celebes (large).	Expanse. Inches.
P. Blumei (Celebes)	5·4
P. Alphenor (Celebes)	4·8
P. Gigon (Celebes)	5·4
P. Deucalion (Celebes)	4·6
P. Agamemnon, var. (Celebes)	4·4
P. Eurypilus (Moluccas)	4·0
P. Telephus (Celebes)	4·3
P. Ægisthus (Moluccas)	4·4
P. Milon (Celebes)	4·4
P. Androcles (Celebes)	4·8
P. Polyphontes (Celebes)	4·6
Leptocircus Ennius (Celebes)	2·0

Closely allied species of Java and the Indian region (small).	Expanse. Inches.
P. Brama	4·0
P. Theseus	3·6
P. Demolion	4·0
P. Macareus	3·7
P. Agamemnon, var.	3 8
P. Jason	3·4
P. Rama	3·2
P. Sarpedon	3·8
P. Antiphates	3·7
P. Diphilus	3·9
L. Meges	1·8

Species inhabiting Amboyna (large).	
Papilio Ulysses	6·1
P. Polydorus	4·9
P. Deiphobus	6·8
P. Gambrisius	6·4
P. Codrus	5·1
Ornithoptera Priamus, (male)	8·3

Allied species of New Guinea and the North Moluccas (smaller).	
P. Autolycus	5 2
P. Telegonus	4·0
P. Leodamas	4·0
P. Deiphontes	5·8
P. Ormenus	5·6
P. Tydeus	6·0
P. Codrus, var. papuensis	4·3
Ornithoptera Poseidon, (male)	7·0

Local variation of Form.—The differences of form are equally clear. Papilio Pammon everywhere on the continent is tailed in both sexes. In Java, Sumatra, and Borneo, the closely allied P. Theseus has a very short tail, or tooth only, in the male, while in the females the tail is retained. Further east, in Celebes and the South Moluccas, the hardly separable P. Alphenor has quite lost the tail in the male, while the female retains it, but in a narrower and less spatulate form. A little further, in Gilolo, P. Nicanor has completely lost the tail in both sexes.

Papilio Agamemnon exhibits a somewhat similar series of changes. In India it is always tailed; in the greater part of the archipelago it has a very short tail; while far east, in New Guinea and the adjacent islands, the tail has almost entirely disappeared.

In the Polydorus-group two species, P. Antiphus and P. Diphilus, inhabiting India and the Indian region, are tailed, while the two which take their place in the Moluccas, New Guinea, and Australia, P. Polydorus and P. Leodamas, are destitute of tail, the species furthest east having lost this ornament the most completely.

Western species, Tailed.	Allied Eastern species not Tailed.
Papilio Pammon (India) ...	P. Thesus (Islands) minute tail.
P. Agamemnon, var. (India)	P. Agamemnon, var. (Islands).
P. Antiphus (India, Java) ...	P. Polydorus (Moluccas).
P. Diphilus (India, Java) ...	P. Leodamas (New Guinea).

The most conspicuous instance of local modification of form, however, is exhibited in the island of Celebes, which in this respect, as in some others, stands alone and isolated in the whole archipelago. Almost every species of Papilio inhabiting Celebes has the wings of a peculiar shape, which distinguishes them at a glance from the allied species of every other island. This peculiarity consists, first, in the upper wings being generally more elongate and falcate; and secondly, in the costa or anterior margin being much more curved, and in most instances exhibiting near the base an abrupt bend or elbow, which in some species is very conspicuous. This peculiarity is visible, not only when the Celebesian species are compared with their small-sized allies of Java and Borneo, but also, and in an almost equal degree, when the large forms of Amboyna and the Moluccas are the objects of comparison, showing that this is quite a distinct phenomenon from the difference of size which has just been pointed out.

In the following Table I have arranged the chief Papilios of Celebes in the order in which they exhibit this characteristic form most prominently.

Papilios of Celebes, having the wings falcate or with abruptly curved costa.	Closely allied Papilios of the surrounding islands, with less falcate wings and slightly curved costa.
1. P. Gigon	P. Demolion (Java).
2. P. Pamphylus	P. Jason (Sumatra).
3. P. Milon	P. Sarpedon (Moluccas, Java).
4. P. Agamemnon, var. ...	P. Agamemnon, var. (Borneo).
5. P. Adamantius	P. Peranthus (Java).
6. P. Ascalaphus	P. Deiphontes (Gilolo).
7. P. Sataspes	P. Helenus (Java).
8. P. Blumei	P. Brama (Sumatra).
9. P. Androcles	P. Antiphates (Borneo).
10. P. Rhesus	P. Aristæus (Moluccas).
11. P. Theseus, var. (male) ...	P. Thesus (male) (Java).
12. P. Codrus, var.	P. Codrus (Moluccas).
13. P. Encelades	P. Leucothoë (Malacca).

It thus appears that every species of Papilio exhibits this peculiar form in a greater or less degree, except one, P. Polyphontes, allied to P. Diphilus of India

and P. Polydorus of the Moluccas. This fact I shall
recur to again, as I think it helps us to understand
something of the causes that may have brought about
the phenomenon we are considering. Neither do the
genera Ornithoptera and Leptocircus exhibit any traces
of this peculiar form. In several other families of
Butterflies this characteristic form reappears in a few
species. In the Pieridæ the following species, all
peculiar to Celebes, exhibit it distinctly :—

1. Picris Eperia ...   ... compared with P. Coronis (Java).
2. Thyca Zebuda...   ...   ,,   ,, Thyca Descombesi
                                                (India).
3. T. Rosenbergii   ...   ,,   ,, T. Hyparete (Java).
4. Tachyris Hombronii ...   ,,   ,, T. Lyncida.
5. T. Lycaste   ...   ...   ,,   ,, T. Lyncida.
6. T. Zarinda   ...   ...   ,,   ,, T. Nero (Malacca).
7. T Ithome   ...   ...   ,,   ,, T. Nephele.
8. Eronia tritæa ...   ...   ,,   ,, Eronia   Valeria
                                                (Java).

9 Iphias Glaucippe, var.   ,,   ,, Iphias   Glaucippe
                                                (Java).

The species of Terias, one or two Pieris, and the genus
Callidryas do not exhibit any perceptible change of
form.

In the other families there are but few similar
examples. The following are all that I can find in my
collection :—

Cethosia Æole   ... compared with Cethosia Biblis (Java).
Eurhinia megalonice   ,,   ,, Eurhinia   Polynice
                                                (Borneo).
Limenitis Limire ...   ,,   ,, Limenitis   Procris
                                                (Java).
Cynthia Arsinoë, var.   ,,   ,, Cynthia Arsinoë (Java,
                                                Sumatra, Borneo)

All these belong to the family of the Nymphalidæ. Many other genera of this family, as Diadema, Adolias, Charaxes, and Cyrestis, as well as the entire families of the Danaidæ, Satyridæ, Lycænidæ, and Hesperidæ, present no examples of this peculiar form of the upper wing in the Celebesian species.

Local variations of Colour.—In Amboyna and Ceram the female of the large and handsome Ornithoptera Helena has the large patch on the hind wings constantly of a pale dull ochre or buff colour, while in the scarcely distinguishable varieties from the adjacent islands of Bouru and New Guinea, it is of a golden yellow, hardly inferior in brilliancy to its colour in the male sex. The female of Ornithoptera Priamus (inhabiting Amboyna and Ceram exclusively) is of a pale dusky brown tint, while in all the allied species the same sex is nearly black with contrasted white markings. As a third example, the female of Papilio Ulysses has the blue colour obscured by dull and dusky tints, while in the closely allied species from the surrounding islands, the females are of almost as brilliant an azure blue as the males. A parallel case to this is the occurrence, in the small islands of Goram, Matabello, Ké, and Aru, of several distinct species of Euplæa and Diadema, having broad bands or patches of white, which do not exist in any of the allied species from the larger islands. These facts seem to indicate some local influence in modifying colour, as unintelligible and almost as remarkable as that which has resulted in the modifications of form previously described.

Remarks on the facts of Local variation.

The facts now brought forward seem to me of the highest interest. We see that almost all the species in two important families of the Lepidoptera (Papilionidæ and Pieridæ) acquire, in a single island, a characteristic modification of form distinguishing them from the allied species and varieties of all the surrounding islands. In other equally extensive families no such change occurs, except in one or two isolated' species. However we may account for these phenomena, or whether we may be quite unable to account for them, they furnish, in my opinion, a strong corroborative testimony in favour of the doctrine of the origin of species by successive small variations; for we have here slight varieties, local races, and undoubted species, all modified in exactly the same manner, indicating plainly a common cause producing identical results. On the generally received theory of the original distinctness and permanence of species, we are met by this difficulty : one portion of these curiously modified forms are admitted to have been produced by variation and some natural action of local conditions; whilst the other portion, differing from the former only in degree, and connected with them by insensible gradations, are said to have possessed this peculiarity of form at their first creation, or to have derived it from unknown causes of a totally distinct nature. Is not the *à priori* evidence in favour of an identity of the causes that have produced such similar results? and have we not a right to call upon our opponents for some proofs of their own doctrine, and for an explanation of its difficulties, instead of their assuming that they are right, and laying upon us the burthen of disproof?

Let us now see if the facts in question do not themselves furnish some clue to their explanation. Mr. Bates has shown that certain groups of butterflies have a defence against insectivorous animals, independent of swiftness of motion. These are generally very abundant, slow, and weak fliers, and are more or less the objects of mimicry by other groups, which thus gain an advantage in a freedom from persecution similar to that enjoyed by those they resemble. Now the only Papilios which have not in Celebes acquired the peculiar form of wing, belong to a group which is imitated both by other species of Papilio and by Moths of the genus Epicopeia. This group is of weak and slow flight; and we may therefore fairly conclude that it possesses some means of defence (probably in a peculiar odour or taste) which saves it from attack. Now the arched costa and falcate form of wing is generally supposed to give increased powers of flight, or, as seems to me more probable, greater facility in making sudden turnings, and thus baffling a pursuer. But the members of the Polydorus-group (to which belongs the only unchanged Celebesian Papilio), being already guarded against attack, have no need of this increased power of wing; and "natural selection" would therefore have no tendency to produce it. The whole family of Danaidæ are in the same position : they are slow and weak fliers ; yet they abound in species and individuals, and are the objects of mimicry. The Satyridæ have also probably a means of protection—perhaps their keeping always near the ground and their generally obscure colours ; while the Lycænidæ and Hesperidæ may find security in their small size and rapid motions. In the extensive family of the Nymphalidæ, however, we find that several of the larger species, of comparatively feeble structure, have their wings modified (Cethosia, Limenitis, Junonia, Cynthia), while the large-bodied powerful species, which have all an excessively rapid flight, have exactly the same form of wing in Celebes as in the other islands. On the whole, therefore, we may say that all the butterflies of rather large size, conspicuous colours, and not very swift flight have been affected in the manner described, while the smaller sized and obscure groups, as well as those which are the objects of mimicry, and also those of exceedingly swift flight have remained unaffected.

It would thus appear as if there must be (or once have been) in the island of Celebes, some peculiar enemy to these larger-sized butterflies which does not exist, or is less abundant, in the surrounding islands. Increased powers of flight, or rapidity of turning, was advantageous in baffling this enemy ; and the peculiar form of wing necessary to give this would be readily acquired by the action of "natural selection" on the slight variations of form that are continually occurring.

Such an enemy one would naturally suppose to be

an insectivorous bird; but it is a remarkable fact that most of the genera of Fly-catchers of Borneo and Java on the one side (Muscipeta, Philentoma,) and of the Moluccas on the other (Monarcha, Rhipidura), are almost entirely absent from Celebes. Their place seems to be supplied by the Caterpillar-catchers (Graucalus, Campephaga, &c.), of which six or seven species are known from Celebes and are very numerous in individuals. We have no positive evidence that these birds pursue butterflies on the wing, but it is highly probable that they do so when other food is scarce. Mr. Bates has suggested to me that the larger Dragonflies (Æshna, &c.) prey upon butterflies; but I did not notice that they were more abundant in Celebes than elsewhere. However this may be, the fauna of Celebes is undoubtedly highly peculiar in every department of which we have any accurate knowledge; and though we may not be able satisfactorily to trace how it has been effected, there can, I think, be little doubt that the singular modification in the wings of so many of the butterflies of that island is an effect of that complicated action and reaction of all living things upon each other in the struggle for existence, which continually tends to readjust disturbed relations, and to bring every species into harmony with the varying conditions of the surrounding universe.

But even the conjectural explanation now given fails us in the other cases of local modification. Why the species of the Western islands should be smaller than those further east,—why those of Amboyna should

N

exceed in size those of Gilolo and New Guinea—why the tailed species of India should begin to lose that appendage in the islands, and retain no trace of it on the borders of the Pacific,—and why, in three separate cases, the females of Amboyna species should be less gaily attired than the corresponding females of the surrounding islands,—are questions which we cannot at present attempt to answer. That they depend, however, on some general principle is certain, because analogous facts have been observed in other parts of the world. Mr. Bates informs me that, in three distinct groups, Papilios which on the Upper Amazon and in most other parts of South America have spotless upper wings obtain pale or white spots at Pará and on the Lower Amazon; and also that the Æneas-group of Papilios never have tails in the equatorial regions and the Amazons valley, but gradually acquire tails in many cases as they range towards the northern or southern tropic. Even in Europe we have somewhat similar facts; for the species and varieties of butterflies peculiar to the island of Sardinia are generally smaller and more deeply coloured than those of the mainland, and the same has recently been shown to be the case with the common tortoiseshell butterfly in the Isle of Man; while Papilio Hospiton, peculiar to the former island, has lost the tail, which is a prominent feature of the closely allied P. Machaon.

Facts of a similar nature to those now brought forward would no doubt be found to occur in other groups of insects, were local faunas carefully studied in relation to those of the surrounding countries; and they seem to indicate that climate and other physical causes have, in some cases, a very powerful effect in modifying specific form and colour, and thus directly aid in producing the endless variety of nature.

Mimicry.

Having fully discussed this subject in the preceding essay, I have only to adduce such illustrations of it, as are furnished by the Eastern Papilionidæ, and to show their bearing upon the phenomena of variation already mentioned. As in America, so in the Old World, species of Danaidæ are the objects which the other families most often imitate. But besides these, some genera of Morphidæ and one section of the genus Papilio are also less frequently copied. Many species of Papilio mimic other species of these three groups so closely that they are undistinguishable when on the wing; and in every case the pairs which resemble each other inhabit the same locality.

The following list exhibits the most important and best marked cases of mimicry which occur among the Papilionidæ of the Malayan region and India :—

Mimickers.	Species mimicked.	Common habitat.
	DANAIDÆ.	
1. Papilio paradoxa (male & female)	Euplœa Midamus (male & female)	Sumatra, &c.
2. P. Caunus	E. Rhadamanthus .	Borneo and Sumatra.
3. P. Thule	Danais sobrina ...	New Guinea.
4. P. Macareus ...	D. Aglaia	Malacca, Java.

Mimickers.	Species mimicked.	Common habitat.
	DANAIDÆ.	
5. Papilio Agestor...	Danais Tytia... ...	Northern India.
6. P. Idæoides ...	Hestia Leuconoë ...	Philippines.
7. P. Delessertii ...	Ideopsis daos ...	Penang.
	MORPHIDÆ.	
8. P. Pandion (female)...	Drusilla bioculata .	New Guinea.

PAPILIO (POLYDORUS- and COON-groups).

9. P. Pammon (Romulus,female)...	Papilio Hector ...	India.
10. P. Theseus, var. (female)	P. Antiphus... ...	Sumatra,Borneo.
11. P. Theseus, var. (female)	P. Diphilus	Sumatra, Java.
12. P. Memnon, var. (Achates,female)	P. Coon...	Sumatra.
13. P. Androgeus,var. (Achates,female)	P. Doubledayi ...	Northern India.
14. P. Œnomaus (female)...	P. Liris	Timor.

We have, therefore, fourteen species or marked varieties of Papilio, which so closely resemble species of other groups in their respective localities, that it is not possible to impute the resemblance to accident. The first two in the list (Papilio paradoxa and P. Caunus) are so exactly like Euplœa Midamus and E. Rhadamanthus on the wing, that although they fly very slowly, I was quite unable to distinguish them. The first is a very interesting case, because the male and female differ considerably, and each mimics the corresponding sex of the Euplœa. A new species of Papilio which I discovered in New Guinea resembles Danais sobrina, from the same country, just as Papilio Marcareus resembles Danais Aglaia in Malacca, and (according to Dr. Horsfield's figure) still more closely in Java. The Indian Papilio Agestor closely imitates Danais Tytia, which has quite a different style of colouring from the preceding; and the extraordinary Papilio Idæoides from the Philippine Islands, must, when on the wing, perfectly resemble the Hestia Leuconoë of the same region, as also does the Papilio Delessertii imitate the Ideopsis daos from Penang. Now in every one of these cases the Papilios are very scarce, while the Danaidæ which they resemble are exceedingly abundant—most of them swarming so as to be a positive nuisance to the collecting entomologist by continually hovering before him when he is in search of newer and more varied captures. Every garden, every roadside, the suburbs of every village are full of them, indicating very clearly that their life is an easy one, and that they are free from persecution by the foes which keep down the population of less favoured races. This superabundant population has been shown by Mr. Bates to be a general characteristic of all American groups and species which are objects of mimicry; and it is interesting to find his observations confirmed by examples on the other side of the globe.

The remarkable genus Drusilla, a group of pale-coloured butterflies, more or less adorned with ocellate spots, is also the object of mimicry by three distinct genera (Melanitis, Hyantis, and Papilio). These insects, like the Danaidæ, are abundant in individuals,

have a very weak and slow flight, and do not seek
concealment, or appear to have any means of protection from insectivorous creatures. It is natural to
conclude, therefore, that they have some hidden property which saves them from attack; and it is easy
to see that when any other insects, by what we call
accidental variation, come more or less remotely to
resemble them, the latter will share to some extent in
their immunity. An extraordinary dimorphic form of
the female of Papilio Ormenus has come to resemble
the Drusillas sufficiently to be taken for one of that
group at a little distance; and it is curious that I captured one of these Papilios in the Aru Islands hovering
along the ground, and settling on it occasionally, just
as it is the habit of the Drusillas to do. The resemblance in this case is only general; but this form of
Papilio varies much, and there is therefore material
for natural selection to act upon, so as ultimately to
produce a copy as exact as in the other cases.

The eastern Papilios allied to Polydorus, Coon,
and Philoxenus, form a natural section of the genus
resembling, in many respects, the Æneas-group of
South America, which they may be said to represent
in the East. Like them, they are forest insects,
have a low and weak flight, and in their favourite
localities are rather abundant in individuals; and like
them, too, they are the objects of mimicry. We may
conclude, therefore, that they possess some hidden
means of protection, which makes it useful to other
insects to be mistaken for them.

The Papilios which resemble them belong to a very distinct section of the genus, in which the sexes differ greatly ; and it is those females only which differ most from the males, and which have already been alluded to as exhibiting instances of dimorphism, which resemble species of the other group.

The resemblance of P. Romulus to P. Hector is, in some specimens, very considerable, and has led to the two species being placed following each other in the British Museum Catalogues and by Mr. E. Doubleday. I have shown, however, that P. Romulus is probably a dimorphic form of the female P. Pammon, and belongs to a distinct section of the genus.

The next pair, Papilio Theseus, and P. Antiphus, have been united as one species both by De Haan and in the British Museum Catalogues. The ordinary variety of P. Theseus found in Java almost as nearly resembles P. Diphilus, inhabiting the same country. The most interesting case, however, is the extreme female form of P. Memnon (figured by Cramer under the name of P. Achates), which has acquired the general form and markings of P. Coon, an insect which differs from the ordinary male P. Memnon, as much as any two species which can be chosen in this extensive and highly varied genus; and, as if to show that this resemblance is not accidental, but is the result of law, when in India we find a species closely allied to P. Coon, but with red instead of yellow spots (P. Doubledayi), the corresponding variety of P. Androgeus (P. Achates, Cramer, 182,

A, B,) has acquired exactly the same peculiarity of having red spots instead of yellow. Lastly, in the island of Timor, the female of P. Œnomaus (a species allied to P. Memnon) resembles so closely P. Liris (one of the Polydorus-group), that the two, which were often seen flying together, could only be distinguished by a minute comparison after being captured.

The last six cases of mimicry are especially instructive, because they seem to indicate one of the processes by which dimorphic forms have been produced. When, as in these cases, one sex differs much from the other, and varies greatly itself, it may happen that occasionally individual variations will occur having a distant resemblance to groups which are the objects of mimicry, and which it is therefore advantageous to resemble. Such a variety will have a better chance of preservation; the individuals possessing it will be multiplied; and their accidental likeness to the favoured group will be rendered permanent by hereditary transmission, and, each successive variation which increases the resemblance being preserved, and all variations departing from the favoured type having less chance of preservation, there will in time result those singular cases of two or more isolated and fixed forms, bound together by that intimate relationship which constitutes them the sexes of a single species. The reason why the females are more subject to this kind of modification than the males is, probably, that their slower flight, when laden with eggs, and their exposure to attack while in the act of depositing their eggs upon leaves, render it especially advantageous for them to have some additional protection. This they at once obtain by acquiring a resemblance to other species which, from whatever cause, enjoy a comparative immunity from persecution.

Concluding remarks on Variation in Lepidoptera.

This summary of the more interesting phenomena of variation presented by the eastern Papilionidæ is, I think, sufficient to substantiate my position, that the Lepidoptera are a group that offer especial facilities for such inquiries; and it will also show that they have undergone an amount of special adaptive modification rarely equalled among the more highly organized animals. And, among the Lepidoptera, the great and pre-eminently tropical families of Papilionidæ and Danaidæ seem to be those in which complicated adaptations to the surrounding organic and inorganic universe have been most completely developed, offering in this respect a striking analogy to the equally extraordinary, though totally different, adaptations which present themselves in the Orchideæ, the only family of plants in which mimicry of other organisms appears to play any important part, and the only one in which cases of conspicuous polymorphism occur; for as such we must class the male, female, and hermaphrodite forms of Catasetum tridentatum, which differ so greatly in form and structure that they were long considered to belong to three distinct genera.

Arrangement and Geographical Distribution of the Malayan Papilionidæ.

Arrangement.—Although the species of Papilionidæ inhabiting the Malayan region are very numerous, they all belong to three out of the nine genera into which the family is divided. One of the remaining genera (Eurycus) is restricted to Australia, and another (Teinopalpus) to the Himalayan Mountains, while no less than four (Parnassius, Doritis, Thais, and Sericinus) are confined to Southern Europe and to the mountain-ranges of the Palæarctic region.

The genera Ornithoptera and Leptocircus are highly characteristic of Malayan entomology, but are uniform in character and of small extent. The genus Papilio, on the other hand, presents a great variety of forms, and is so richly represented in the Malay Islands, that more than one-fourth of all the known species are found there. It becomes necessary, therefore, to divide this genus into natural groups before we can successfully study its geographical distribution.

Owing principally to Dr. Horsfield's observations in Java, we are acquainted with a considerable number of the larvæ of Papilios ; and these furnish good characters for the primary division of the genus into natural groups. The manner in which the hinder wings are plaited or folded back at the abdominal margin, the size of the anal valves, the structure of the antennæ, and the form of the wings are also of much service, as well as the character of the flight and the style of

colouration. Using these characters, I divide the
Malayan Papilios into four sections, and seventeen
groups, as follows:—

Genus ORNITHOPTERA.

    a. Priamus-group.
    c. Brookeanus-group. } Black and green.
    b. Pompeus-group.   Black and yellow.

Genus PAPILIO.

A. Larvæ short, thick, with numerous fleshy tubercles;
    of a purplish colour.

    a. Nox-group. Abdominal fold in male very large;
    anal valves small, but swollen; antennæ mode-
    rate; wings entire, or tailed; includes the Indian
    Philoxenus-group.

    b. Coon-group. Abdominal fold in male small; anal
    valves small, but swollen; antennæ moderate;
    wings tailed.

    c. Polydorus-group. Abdominal fold in male small,
    or none; anal valves small or obsolete, hairy;
    wings tailed or entire.

B. Larvæ with third segment swollen, transversely or
    obliquely banded; pupa much bent. Imago with
    abdominal margin in male plaited, but not re-
    flexed; body weak; antennæ long; wings much
    dilated, often tailed.

    d. Ulysses-group.

    e. Peranthus-group. } Protenor-group (Indian) is
    f. Memnon-group. } somewhat intermediate be-
    tween these, and is nearest
    to the Nox-group.

    g. Helenus-group.
    h. Erectheus-group.
    i. Pammon-group.
    k. Demolion-group.

C. Larvæ subcylindrical, variously coloured. Imago with
    abdominal margin in male plaited, but not re-
    flexed; body weak; antennæ short, with a thick
    curved club; wings entire.

> l. Erithonius-group. Sexes alike, larva and pupa
> something like those of P. Demolion.
> m. Paradoxa-group. Sexes different.
> n. Dissimilis - group.   Sexes alike ; larva bright -
> coloured; pupa straight, cylindric.

D. Larvæ elongate, attenuate behind, and often bifid, with
lateral and oblique pale stripes, green. Imago
with the abdominal margin in male reflexed,
woolly or hairy within; anal valves small, hairy;
antennæ short, stout; body stout.

> o. Macareus-group. Hind wings entire.
> p. Antiphates-group. Hind wings much tailed (swal-
> low-tails).
> q. Eurypylus-group. Hind wings elongate or tailed.

Genus LEPTOCIRCUS.

Making, in all, twenty· distinct groups of Malayan
Papilionidæ.

The first section of the genus Papilio (A) comprises
insects which, though differing considerably in structure, having much general resemblance. They all have
a weak, low flight, frequent the most luxuriant forest-
districts, seem to love the shade, and are the objects
of mimicry by other Papilios.

Section B consists of weak-bodied, large-winged insects, with an irregular wavering flight, and which,
when resting on foliage, often expand the wings, which
the species of the other sections rarely or never do.
They are the most conspicuous and striking of eastern
Butterflies.

Section C consists of much weaker and slower-flying
insects, often resembling in their flight, as well as in
their colours, species of Danaidæ.

Section D contains the strongest-bodied and most swift-flying of the genus. They love sunlight, and frequent the borders of streams and the edges of puddles, where they gather together in swarms consisting of several species, greedily sucking up the moisture, and, when disturbed, circling round in the air, or flying high and with great strength and rapidity.

Geographical Distribution.—One hundred and thirty species of Malayan Papilionidæ are now known within the district extending from the Malay peninsula, on the north-west, to Woodlark Island, near New Guinea, on the south-east.

The exceeding richness of the Malayan region in these fine insects is seen by comparing the number of species found in the different tropical regions of the earth. From all Africa only 33 species of Papilio are known ; but as several are still undescribed in collections, we may raise their number to about 40. In all tropical Asia there are at present described only 65 species, and I have seen in collections but two or three which have not yet been named. In South America, south of Panama, there are 150 species, or about one-seventh more than are yet known from the Malayan region ; but the area of the two countries is very different ; for while South America (even excluding Patagonia) contains 5,000,000 square miles, a line encircling the whole of the Malayan islands would only include an area of 2,700,000 square miles, of which the land-area would be about 1,000,000 square miles. This superior richness is partly real and partly apparent. The breaking up of a district into small isolated portions, as in an archipelago, seems highly favourable to the segregation and perpetuation of local peculiarities in certain groups; so that a species which on a continent might have a wide range, and whose local forms, if any, would be so connected together that it would be impossible to separate them, may become by isolation reduced to a number of such clearly defined and constant forms that we are obliged to count them as species. From this point of view, therefore, the greater proportionate number of Malayan species may be considered as apparent only. Its true superiority is shown, on the other hand, by the possession of three genera and twenty groups of Papilionidæ against a single genus and eight groups in South America, and also by the much greater average size of the Malayan species. In most other families, however, the reverse is the case, the South American Nymphalidæ, Satyridæ, and Erycinidæ far surpassing those of the East in number, variety, and beauty.

The following list, exhibiting the range and distribution of each group, will enable us to study more easily their internal and external relations.

Range of the Groups of Malayan Papilionidæ.
Ornithoptera.

    1. Priamus - group.   Moluccas to Woodlark
        Island ...   . 5 species.
    2. Pompeus - group.   Himalayas to New
        Guinea, (Celebes, maximum)   11   „
    3. Brookeana-group. Sumatra and Borneo .. 1   „

Papilio.

   4. Nox-group. North India, Java, and Philippines ... ... 5 species.
   5. Coon-group. North India to Java... ... 2 „
   6. Polydorus-group. India to New Guinea and Pacific ... ... 7 „
   7. Ulysses-group. Celebes to New Caledonia 4 „
   8. Peranthus - group. India to Timor and Moluccas (India, maximum) ... ... 9 „
   9. Memnon-group. India to Timor and Moluccas (Java, maximum) ... ... 10 „
  10. Helenus-group. Africa and India to New Guinea ... ... 11 „
  11. Pammon-group. India to Pacific and Australia ... ... 9 „
  12. Erectheus-group. Celebes to Australia ... 8 „
  13. Demolion-group. India to Celebes ... 2 „
  14. Erithonius-group. Africa, India, Australia 1 „
  15. Paradoxa-group. India to Java (Borneo, maximum) ... ... 5 „
  16. Dissimilis-group. India to Timor (India, maximum) ... ... 2 „
  17. Macareus-group. India to New Guinea ... 10 „
  18. Antiphates-group. Widely distributed ... 8 „
  19. Eurypylus-group. India to Australia ,.. 15 „

Leptocircus.

  20 Leptocircus-group. India to Celebes ... 4 „

This Table shows the great affinity of the Malayan with the Indian Papilionidæ, only three out of the twenty groups ranging beyond, into Africa, Europe, or America. The limitation of groups to the Indo-Malayan or Austro-Malayan divisions of the archipelago, which is so well marked in the higher animals, is much less conspicuous in insects, but is shown in some degree by the Papilionidæ. The following groups

are either almost or entirely restricted to one portion
of the archipelago :—

Indo-Malayan Region.	*Austro-Malayan Region.*
Nox-group.	Priamus-group.
Coon-group.	Ulysses-group.
Macareus-group (nearly).	Erechtheus-group.
Paradoxa-group.	
Dissimilis-group (nearly).	
Brookeanus-group.	
LEPTOCIRCUS (genus).	

The remaining groups, which range over the whole
archipelago, are, in many cases, insects of very powerful flight, or they frequent open places and the sea-
beach, and are thus more likely to get blown from
island to island. The fact that three such characteristic groups as those of Priamus, Ulysses, and Erechtheus are strictly limited to the Australian region of
the archipelago, while five other groups are with equal
strictness confined to the Indian region, is a strong
corroboration of that division which has been founded
almost entirely on the distribution of Mammalia and
Birds.

If the various Malayan islands have undergone
recent changes of level, and if any of them have been
more closely united within the period of existing
species than they are now, we may expect to find
indications of such changes in community of species
between islands now widely separated ; while those
islands which have long remained isolated would have
had time to acquire peculiar forms by a slow and
natural process of modification.

An examination of the relations of the species of the adjacent islands, will thus enable us to correct opinions formed from a mere consideration of their relative positions. For example, looking at a map of the archipelago, it is almost impossible to avoid the idea that Java and Sumatra have been recently united; their present proximity is so great, and they have such an obvious resemblance in their volcanic structure. Yet there can be little doubt that this opinion is erroneous, and that Sumatra has had a more recent and more intimate connexion with Borneo than it has had with Java. This is strikingly shown by the mammals of these islands—very few of the species of Java and Sumatra being identical, while a considerable number are common to Sumatra and Borneo. The birds show a somewhat similar relationship; and we shall find that the distribution of the Papilionidæ tells exactly the same tale. Thus :—

Sumatra has... 21 species	}	20 sp. common to both islands;
Borneo ,, ... 30 ,,	
Sumatra ,, ... 21 ,,	}	11 sp. common to both islands ;
Java ,, ... 28 ,,	
Borneo ,, ... 30 ,,	}	20 sp. common to both islands ;
Java ,, ... 28 ,,	

showing that both Sumatra and Java have a much closer relationship to Borneo than they have to each other—a most singular and interesting result, when we consider the wide separation of Borneo from them both, and its very different structure. The evidence furnished by a single group of insects would have had

o

but little weight on a point of such magnitude if standing alone; but coming as it does to confirm deductions drawn from whole classes of the higher animals, it must be admitted to have considerable value.

We may determine in a similar manner the relations of the different Papuan Islands to New Guinea. Of thirteen species of Papilionidæ obtained in the Aru Islands, six were also found in New Guinea, and seven not. Of nine species obtained at Waigiou, six were New Guinea, and three not. The five species found at Mysol were all New Guinea species. Mysol, therefore, has closer relations to New Guinea than the other islands; and this is corroborated by the distribution of the birds, of which I will only now give one instance. The Paradise Bird found in Mysol is the common New Guinea species, while the Aru Islands and Waigiou have each a species peculiar to themselves.

The large island of Borneo, which contains more species of Papilionidæ than any other in the archipelago, has nevertheless only three peculiar to itself; and it is quite possible, and even probable, that one of these may be found in Sumatra or Java. The last-named island has also three species peculiar to it; Sumatra has not one, and the peninsula of Malacca only two. The identity of species is even greater than in birds or in most other groups of insects, and points very strongly to a recent connexion of the whole with each other and the continent.

Remarkable Peculiarities of the Island of Celebes.

If we now pass to the next island (Celebes), separated from those last mentioned by a strait not wider than that which divides them from each other, we have a striking contrast; for with a total number of species less than either Borneo or Java, no fewer than eighteen are absolutely restricted to it. Further east, the large islands of Ceram and New Guinea have only three species peculiar to each, and Timor has five. We shall have to look, not to single islands, but to whole groups, in order to obtain an amount of individuality comparable with that of Celebes. For example, the extensive group comprising the large islands of Java, Borneo, and Sumatra, with the peninsula of Malacca, possessing altogether 48 species, has about 24, or just half, peculiar to it; the numerous group of the Philippines possess 22 species, of which 17 are peculiar; the seven chief islands of the Moluccas have 27, of which 12 are peculiar; and the whole of the Papuan Islands, with an equal number of species, have 17 peculiar. Comparable with the most isolated of these groups is Celebes, with its 24 species, of which the large proportion of 18 are peculiar. We see, therefore, that the opinion I have elsewhere expressed, of the high degree of isolation and the remarkable distinctive features of this interesting island, is fully borne out by the examination of this conspicuous family of insects. A single straggling island with a few small satellites, it is zoologically of equal

importance with extensive groups of islands many
times as large as itself; and standing in the very centre
of the archipelago, surrounded on every side with islets
connecting it with the larger groups, and which seem
to afford the greatest facilities for the migration and
intercommunication of their respective productions, it
yet stands out conspicuous with a character of its own
in every department of nature, and presents peculiarities which are, I believe, without a parallel in any
similar locality on the globe.

Briefly to summarize these peculiarities, Celebes
possesses three genera of mammals (out of the very
small number which inhabit it) which are of singular
and isolated forms, viz., Cynopithecus, a tailless Ape
allied to the Baboons; Anoa, a straight-horned Antelope of obscure affinities, but quite unlike anything
else in the whole archipelago or in India: and Babi-
rusa, an altogether abnormal wild Pig. With a rather
limited bird population, Celebes has an immense preponderance of species confined to it, and has also six
remarkable genera (Meropogon, Ceycopsis, Strepto-
citta, Enodes, Scissirostrum, and Megacephalon) entirely restricted to its narrow limits, as well as two
others (Prioniturus and Basilornis) which only range
to a single island beyond it.

Mr. Smith's elaborate tables of the distribution of
Malayan Hymenoptera (see " Proc. Linn. Soc." Zool.
vol. vii.) show that out of the large number of 301
species collected in Celebes, 190 (or nearly two-thirds)
are absolutely restricted to it, although Borneo on one

side, and the various islands of the Moluccas on the other, were equally well explored by me; and no less than twelve of the genera are not found in any other island of the archipelago. I have shown in the present essay that, in the Papilionidæ, it has far more species of its own than any other island, and a greater proportion of peculiar species than many of the large groups of islands in the archipelago—and that it gives to a large number of the species and varieties which inhabit it, 1st, an increase of size, and, 2nd, a peculiar modification in the form of the wings, which stamp upon the most dissimilar insects a mark distinctive of their common birth-place.

What, I would ask, are we to do with phenomena such as these? Are we to rest content with that very simple, but at the same time very unsatisfying explanation, that all these insects and other animals were created exactly *as* they are, and originally placed exactly *where* they are, by the inscrutable will of their Creator, and that we have nothing to do but to register the facts and wonder? Was this single island selected for a fantastic display of creative power, merely to excite a childlike and unreasoning admiration? Is all this appearance of gradual modification by the action of natural causes—a modification the successive steps of which we can almost trace—all delusive? Is this harmony between the most diverse groups, all presenting analogous phenomena, and indicating a dependence upon physical changes of which we have independent evidence, all false testimony? If I could think so, the study of nature would have lost for me its greatest charm. I should feel as would the geologist, if you could convince him that his interpretation of the earth's past history was all a delusion—that strata were never formed in the primeval ocean, and that the fossils he so carefully collects and studies are no true record of a former living world, but were all created just as they now are, and in the rocks where he now finds them.

I must here express my own belief that none of these phenomena, however apparently isolated or insignificant, can ever stand alone—that not the wing of a butterfly can change in form or vary in colour, except in harmony with, and as a part of the grand march of nature. I believe, therefore, that all the curious phenomena I have just recapitulated, are immediately dependent on the last series of changes, organic and inorganic, in these regions ; and as the phenomena presented by the island of Celebes differ from those of all the surrounding islands, it can, I conceive, only be because the past history of Celebes has been, to some extent, unique and different from theirs. We must have much more evidence to determine exactly in what that difference has consisted. At present, I only see my way clear to one deduction, viz., that Celebes represents one of the oldest parts of the archipelago ; that it has been formerly more completely isolated both from India and from Australia than it is now, and that amid all the mutations it has undergone, a relic or substratum of the fauna and flora of some more ancient land has been here preserved to us.

It is only since my return home, and since I have been able to compare the productions of Celebes side by side with those of the surrounding islands, that I have been fully impressed with their peculiarity, and the great interest that attaches to them. The plants and the reptiles are still almost unknown; and it is to be hoped that some enterprising naturalist may soon devote himself to their study. The geology of the country would also be well worth exploring, and its newer fossils would be of especial interest as elucidating the changes which have led to its present anomalous condition. This island stands, as it were, upon the boundary-line between two worlds. On one side is that ancient Australian fauna, which preserves to the present day the facies of an early geological epoch; on the other is the rich and varied fauna of Asia, which seems to contain, in every class and order, the most perfect and highly organised animals. Celebes has relations to both, yet strictly belongs to neither: it possesses characteristics which are altogether its own; and I am convinced that no single island upon the globe would so well repay a careful and detailed research into its past and present history.

Concluding Remarks.

In writing this essay it has been my object to show how much may, under favourable circumstances, be learnt by the study of what may be termed the external physiology of a small group of animals, inhabiting a limited district. This branch of natural history had received little attention till Mr. Darwin showed how important an adjunct it may become towards a true interpretation of the history of organized beings, and attracted towards it some small share of that research which had before been almost exclusively devoted to internal structure and physiology. The nature of species, the laws of variation, the mysterious influence of locality on both form and colour, the phenomena of dimorphism and of mimicry, the modifying influence of sex, the general laws of geographical distribution, and the interpretation of past changes of the earth's surface, have all been more or less fully illustrated by the very limited group of the Malayan Papilionidæ; while, at the same time, the deductions drawn therefrom have been shown to be supported by analogous facts, occurring in other and often widely-separated groups of animals.

V.

ON INSTINCT IN MAN AND ANIMALS.

THE most perfect and most striking examples of what is termed instinct, those in which reason or observation appear to have the least influence, and which seem to imply the possession of faculties farthest removed from our own, are to be found among insects. The marvellous constructive powers of bees and wasps, the social economy of ants, the careful provision for the safety of a progeny they are never to see manifested by many beetles and flies, and the curious preparations for the pupa state by the larvæ of butterflies and moths, are typical examples of this faculty, and are supposed to be conclusive as to the existence of some power or intelligence, very different from that which we derive from our senses or from our reason.

How Instinct may be best Studied.

Whatever we may define instinct to be, it is evidently some form of mental manifestation, and as we can only judge of mind by the analogy of our own mental functions and by observation of the results of mental action in other men and in animals, it is incumbent on us, first, to study and endeavour to comprehend the minds of infants, of savage men, and of animals not very far removed from ourselves, before we pronounce positively as to the nature of the mental operations in creatures so radically different from us as insects. We have not yet even been able to ascertain what are the senses they possess, or what relation their powers of seeing, hearing, and feeling have to ours. Their sight may far exceed ours both in delicacy and in range, and may .possibly give them knowledge of the internal constitution of bodies analogous to that which we obtain by the spectroscope; and that their visual organs do possess some powers which ours do not, is indicated by the extraordinary crystalline rods radiating from the optic ganglion to the facets of the compound eye, which rods vary in form and thickness in different parts of their length, and possess distinctive characters in each group of insects. This complex apparatus, so different from anything in the eyes of vertebrates, may subserve some function quite inconceivable by us, as well as that which we know as vision. There is reason to believe that insects appreciate sounds of extreme delicacy, and it is supposed that certain minute organs, plentifully supplied with nerves, and situated in the subcostal vein of the wing in most insects, are the organs of hearing. But besides these, the Orthoptera (such as grasshoppers, &c.) have what are supposed to be ears on their fore legs, and Mr. Lowne believes that the little stalked balls, which are the sole remnants of the hind wings in flies, are also organs of hearing or of some analogous sense. In flies, too, the third joint of the antennæ contains thousands of nerve-fibres, which terminate in small open cells, and this Mr. Lowne believes to be the organ of smell, or of some other, perhaps new, sense. It is quite evident, therefore, that insects may possess senses which give them a knowledge of that which we can never perceive, and enable them to perform acts which to us are incomprehensible. In the midst of this complete ignorance of their faculties and inner nature, is it wise for us to judge so boldly of their powers by a comparison with our own? How can we pretend to fathom the profound mystery of their mental nature, and decide what, and how much, they can perceive or remember, reason or reflect! To leap at one bound from our own consciousness to that of an insect's, is as unreasonable and absurd as if, with a pretty good knowledge of the multiplication table, we were to go straight to the study of the calculus of functions, or as if our comparative anatomists should pass from the study of man's bony structure to that of the fish, and, without any knowledge of the numerous intermediate forms, were to attempt to determine the homologies between these distant types of vertebrata. In such a case would not error be inevitable, and would not continued study in the same direction only render the erroneous conclusions more ingrained and more irremovable.

Definition of Instinct.

Before going further into this subject, we must

determine what we mean by the term instinct. It
has been variously defined as—" disposition operating
without the aid of instruction or experience," " a
mental power totally independent of organization," or
"a power enabling an animal to do that which, in those
things man can do, results from a chain of reasoning,
and in things which man cannot do, is not to be explained by any efforts of the intellectual faculties."
We find, too, that the word instinct is very frequently
applied to acts which are evidently the result either
of organization or of habit. The colt or calf is said
to walk instinctively, almost as soon as it is born;
but this is solely due to its organization, which renders walking both possible and pleasurable to it. So
we are said instinctively to hold out our hands to
save ourselves from falling, but this is an acquired
habit, which the infant does not possess. It appears
to me that instinct should be defined as—" the per-
formance by an animal of complex acts, absolutely
without instruction or previously-acquired knowledge."
Thus, acts are said to be performed by birds in building their nests, by bees in constructing their cells,
and by many insects in providing for the future wants
of themselves or their progeny, without ever having
seen such acts performed by others, and without any
knowledge of why they perform them themselves.
This is expressed by the very common term " blind
instinct." But we have here a number of assertions
of matters of fact, which, strange to say, have never
been proved to be facts at all. They are thought to

be so self-evident that they may be taken for granted. No one has ever yet obtained the eggs of some bird which builds an elaborate nest, hatched these eggs by steam or under a quite distinct parent, placed them afterwards in an extensive aviary or covered garden, where the situation and the materials of a nest similar to that of the parent birds may be found, and then seen what kind of nest these birds would build. If under these rigorous conditions they choose the same materials, the same situation, and construct the nest in the same way and as per'ectly as their parents did, instinct would be proved in their case; now it is only assumed, and assumed, as I shall show further on, without any sufficient reason. So, no one has ever carefully taken the pupæ of a hive of bees out of the comb, removed them from the presence of other bees, and loosed them in a large conservatory with plenty of flowers and food, and observed what kind of cells they would construct. But till this is done, no one can say that bees build without instruction, no one can say that with every new swarm there are no bees older than those of the same year, who may be the teachers in forming the new comb. Now, in a scientific inquiry, a point which can be proved should not be assumed, and a totally unknown power should not be brought in to explain facts, when known powers may be sufficient. For both these reasons I decline to accept the theory of instinct in any case where all other possible modes of explanation have not been exhausted.

Does Man possess Instincts.

Many of the upholders of the instinctive theory maintain, that man has instincts exactly of the same nature as those of animals, but more or less liable to be obscured by his reasoning powers; and as this is a case more open to our observation than any other, I will devote a few pages to its consideration. Infants are said to suck by instinct, and afterwards to walk by the same power, while in adult man the most prominent case of instinct is supposed to be, the powers possessed by savage races to find their way across a trackless and previously unknown wilderness. Let us take first the case of the infant's sucking. It is sometimes absurdly stated that the new-born infant " seeks the breast," and this is held to be a wonderful proof of instinct. No doubt it would be if true, but unfortunately for the theory it is totally false, as every nurse and medical man can testify. Still, the child undoubtedly sucks without teaching, but this is one of those *simple* acts dependent upon organization, which cannot properly be termed instinct, any more than breathing or muscular motion. Any object of suitable size in the mouth of an infant excites the nerves and muscles so as to produce the act of suction, and when at a little later period, the will comes into play, the pleasurable sensations consequent on the act lead to its continuance. So, walking is evidently dependent on the arrangement of the bones and joints, and the pleasurable exertion of the muscles, which lead to the vertical posture becoming gradually the most agreeable one ; and there can be little doubt that an infant would learn of itself to walk, even if suckled by a wild beast.

How Indians travel through unknown and trackless Forests.

Let us now consider the fact, of Indians finding their way through forests they have never traversed before. This is much misunderstood, for I believe it is only performed under such special conditions, as at once to show that instinct has nothing to do with it. A savage, it is true, can find his way through his native forests in a direction in which he has never traversed them before; but this is because from infancy he has been used to wander in them, and to find his way by indications which he has observed himself or learnt from others. Savages make long journeys in many directions, and, their whole faculties being directed to the subject, they gain a wide and accurate knowledge of the topography, not only of their own district, but of all the regions round about. Every one who has travelled in a new direction communicates his knowledge to those who have travelled less. and descriptions of routes and localities, and minute incidents of travel, form one of the main staples of conversation round the evening fire. Every wanderer or captive from another tribe adds to the store of information, and as the very existence of individuals and of whole families and tribes, depends upon the completeness of this knowledge, all the acute perceptive faculties of the adult savage are devoted to acquiring and perfecting it. The good hunter or warrior thus comes to know the bearing of every hill and mountain range, the directions and junctions of all the streams, the situation of each tract characterized by peculiar vegetation, not only within the area he has himself traversed, but for perhaps a hundred miles around it. His acute observation enables him to detect the slightest undulations of the surface, the various changes of subsoil and alterations in the character of the vegetation, that would be quite imperceptible to a stranger. His eye is always open to the direction in which he is going; the mossy side of trees, the presence of certain plants under the shade of rocks, the morning and evening flight of birds, are to him indications of direction, almost as sure as the sun in the heavens. Now, if such a savage is required to find his way across this country in a direction in which he has never been before, he is quite equal to the task. By however circuitous a route he has come to the point he is to start from, he has observed all the bearings and distances so well, that he knows pretty nearly where he is, the direction of his own home and that of the place he is required to go to. He starts towards it, and knows that by a certain time he must cross an upland or a river, that the streams should flow in a certain direction, and that he should cross some of them at a certain distance from their sources. The nature of the soil throughout the whole region is known to him, as well as all the great features of the vegetation. As he approaches any tract of country he has been in or near before, many minute indications guide him, but he observes them so cautiously that his white companions cannot perceive by what he has directed his course. Every now and then he slightly changes his direction, but he is never confused, never loses himself, for he always feels at home; till at last he arrives at a well-known country, and directs his course so as to reach the exact spot desired. To the Europeans whom he guides, he seems to have come without trouble, without any special observation, and in a nearly straight unchanging course. They are astonished, and ask if he has ever been the same route before, and when he answers " No," conclude that some unerring instinct could alone have guided him. But take this same man into another country very similar to his own, but with other streams and hills, another kind of soil, with a somewhat different vegetation and animal life; and after bringing him by a circuitous route to a given point, ask him to return to his starting place, by a straight line of fifty miles through the forest, and he will certainly decline to attempt it, or, attempting it, will more or less completely fail. His supposed instinct does not act out of his own country.

A savage, even in a new country, has, however, undoubted advantages, from his familiarity with forest life, his entire fearlessness of being lost, his accurate perception of direction and of distance, and he is thus able very soon to acquire a knowledge of the district that seems marvellous to a civilized man; but my own observation of savages in forest countries has convinced me, that they find their way by the use of no other faculties than those which we ourselves possess. It appears to me, therefore, that to call in the aid of a new and mysterious power to account for savages being able to do that which, under similar conditions, we could almost all of us perform, although perhaps less perfectly, is almost ludicrously unnecessary.

In the next essay I shall attempt to show, that much of what has been attributed to instinct in birds, can be also very well explained by crediting them with those faculties of observation, memory, and imitation, and with that limited amount of reason, which they undoubtedly exhibit.

VI.

THE PHILOSOPHY OF BIRDS' NESTS.

Instinct or Reason in the Construction of Birds' Nests.

BIRDS, we are told, build their nests by *instinct*, while man constructs his dwelling by the exercise of *reason*. Birds never change, but continue to build for ever on the self-same plan; man alters and improves his houses continually. Reason advances; instinct is stationary.

This doctrine is so very general that it may almost be said to be universally adopted. Men who agree on nothing else, accept this as a good explanation of the facts. Philosophers and poets, metaphysicians and divines, naturalists and the general public, not only agree in believing this to be probable, but even adopt it as a sort of axiom that is so self-evident as to need no proof, and use it as the very foundation of their speculations on instinct and reason. A belief so general, one would think, must rest on indisputable facts, and be a logical deduction from them. Yet I have come to the conclusion that not only is it very doubtful, but absolutely erroneous; that it not only deviates widely from the truth, but is in almost every particular exactly opposed to it. I believe, in short, that birds do *not* build their nests by instinct; that man does *not* con-

P 2

struct his dwelling by reason ; that birds do change and improve when affected by the same causes that make men do so; and that mankind neither alter nor improve when they exist under conditions similar to those which are almost universal among birds.

Do Men build by Reason or by Imitation?

Let us first consider the theory of reason, as alone determining the domestic architecture of the human race.   Man, as a reasonable animal, it is said, continually alters and improves his dwelling.   This I entirely deny.   As a rule, he neither alters nor improves, any more than the birds do.   What have the houses of most savage tribes improved from, each as invariable as the nest of a species of bird?   The tents of the Arab are the same now as they were two or three thousand years ago, and the mud villages of Egypt can scarcely have improved since the time of the Pharoahs.   The palm-leaf huts and hovels of the various tribes of South America and the Malay Archipelago, what have they improved from since those regions were first inhabited? The Patagonian's rude shelter of leaves, the hollowed bank of the South African Earthmen, we cannot even conceive to have been ever inferior to what they now are.   Even nearer home, the Irish turf cabin and the Highland stone shelty can hardly have advanced much during the last two thousand years.   Now, no one imputes this stationary condition of domestic architecture among these savage tribes to instinct, but to simple imitation from one generation to another, and the absence of any sufficiently powerful stimulus to change or improvement. No one imagines that if an infant Arab could be transferred to Patagonia or to the Highlands, it would, when it grew up, astonish its foster-parents by constructing a tent of skins. On the other hand, it is quite clear that physical conditions, combined with the degree of civilization arrived at, almost necessitate certain types of structure. The turf, or stones, or snow—the palm-leaves, bamboo, or branches, which are the materials of houses in various countries, are used because nothing else is so readily to be obtained. The Egyptian peasant has none of these, not even wood. What, then, can he use but mud? In tropical forest-countries, the bamboo and the broad palm-leaves are the natural material for houses, and the form and mode of structure will be decided in part by the nature of the country, whether hot or cool, whether swampy or dry, whether rocky or plain, whether frequented by wild beasts, or whether subject to the attacks of enemies. When once a particular mode of building has been adopted, and has become confirmed by habit and by hereditary custom, it will be long retained, even when its utility has been lost through changed conditions, or through migration into a very different region. As a general rule, throughout the whole continent of America, native houses are built directly upon the ground—strength and security being given by thickening the low walls and the roof. In almost the whole of the Malay Islands, on the contrary, the houses are raised on posts, often to a great height, with an open bamboo floor; and the whole structure is exceedingly slight and thin. Now, what can be the reason of this remarkable difference between countries, many parts of which are strikingly similar in physical conditions, natural productions, and the state of civilization of their inhabitants? We appear to have some clue to it in the supposed origin and migrations of their respective populations. The indigenes of tropical America are believed to have immigrated from the north—from a country where the winters are severe, and raised houses with open floors would be hardly habitable. They moved southwards by land along the mountain ranges and uplands, and in an altered climate continued the mode of construction of their forefathers, modified only by the new materials they met with. By minute observations of the Indians of the Amazon Valley, Mr. Bates arrived at the conclusion that they were comparatively recent immigrants from a colder climate. He says:—"No one could live long among the Indians of the Upper Amazon without being struck with their constitutional dislike to the heat. . . Their skin is hot to the touch, and they perspire little. . . They are restless and discontented in hot, dry weather, but cheerful on cool days, when the rain is pouring down their naked backs." And, after giving many other details, he concludes, " How different all this is with the Negro, the true child of tropical climes! The impression gradually forced itself on my mind that the Red Indian lives as an immigrant or stranger in these hot regions, and that his constitution was not originally adapted, and has not since become perfectly adapted, to the climate."

The Malay races, on the other hand, are no doubt very ancient inhabitants of the hottest regions, and are particularly addicted to forming their first settlements at the mouths of rivers or creeks, or in landlocked bays and inlets. They are a pre-eminently maritime or semi-aquatic people, to whom a canoe is a necessary of life, and who will never travel by land if they can do so by water. In accordance with these tastes, they have built their houses on posts in the water, after the manner of the lake-dwellers of ancient Europe ; and this mode of construction has become so confirmed, that even those tribes who have spread far into the interior, on dry plains and rocky mountains, continue to build in exactly the same manner, and find safety in the height to which they elevate their dwellings above the ground.

Why does each Bird build a peculiar kind of Nest?

These general characteristics of the abode of savage man will be found to be exactly paralleled by the nests of birds. Each species uses the materials it can most readily obtain, and builds in situations most congenial to its habits. The wren, for example, frequenting hedgerows and low thickets, builds its nest generally of *moss*, a material always found where it lives, and among which it probably obtains much of its insect food ; but it varies sometimes, using hay or feathers when these are at hand. Rooks dig in pastures and ploughed fields for grubs, and in doing so must continually encounter *roots* and *fibres*. These are used to line its nest. What more natural ! The crow feeding on carrion, dead rabbits, and lambs, and frequenting sheep-walks and warrens, chooses *fur* and *wool* to line its nest. The lark frequents cultivated fields, and makes its nest, on the ground, of grass lined with *horsehair*—materials the most easy to meet with, and the best adapted to its needs. The kingfisher makes its nest of the *bones* of the fish which it has eaten. Swallows use clay and mud from the margins of the ponds and rivers over which they find their insect food. The materials of birds' nests, like those used by savage man for his house, are, then, those which come first to hand; and it certainly requires no more special instinct to select them in one case than in the other.

But, it will be said, it is not so much the materials as the form and structure of nests, that vary so much, and are so wonderfully adapted to the wants and habits of each species; how are these to be accounted for except by instinct? I reply, they may be in a great measure explained by the general habits of the species, the nature of the tools they have to work with, and the materials they can most easily obtain, with the very simplest adaptations of means to an end, quite within the mental capacities of birds. The delicacy and perfection of the nest will bear a direct relation to the size of the bird, its structure and habits. That of the wren or the humming-bird is perhaps not finer or more beautiful in proportion than that of the blackbird, the magpie, or the crow. The wren, having a slender beak, long legs, and great activity, is able with great ease to form a well-woven nest of the finest materials, and places it in thickets and hedgerows which it frequents in its search for food. The titmouse, haunting fruit-trees and walls, and searching in cracks and crannies for insects, is naturally led to build in holes where it has shelter and security; while its great activity, and the perfection of its tools (bill and feet), enable it readily to form a beautiful receptable for its eggs and young. Pigeons having heavy bodies and weak feet and bills (imperfect tools for forming a delicate structure) build rude, flat nests of sticks, laid across strong branches which will bear their weight and that of their bulky young. They can do no better. The Caprimulgidæ have the most imperfect tools of all, feet that will not support them except on a flat surface (for they cannot truly perch) and a bill excessively broad, short, and weak, and almost hidden by feathers and bristles. They cannot build a nest of twigs or fibres, hair or moss, like other birds, and they therefore generally dispense with one altogether, laying their eggs on the bare ground, or on the stump or flat limb of a tree. The clumsy hooked bills, short necks and feet, and heavy bodies of Parrots, render them quite incapable of building a nest like most other birds. They cannot climb up a branch without using both bill and feet; they cannot even turn round on a perch without holding on with their bill. How, then, could they inlay, or weave, or twist the materials of a nest? Consequently, they all lay in holes of trees, the tops of rotten stumps, or in deserted ants' nests, the soft materials of which they can easily hollow out.

Many terns and sandpipers lay their eggs on the bare sand of the sea-shore, and no doubt the Duke of Argyll is correct when he says, that the cause of this habit is not that they are unable to form a nest, but that, in such situations, any nest would be conspicuous and lead to the discovery of the eggs. The choice of *place* is, however, evidently determined by the habits of the birds, who, in their daily search for food, are continually roaming over extensive tide-washed flats. Gulls vary considerably in their mode of nesting, but it is always in accordance with their structure and habits. The situation is either on a bare rock or on ledges of sea-cliffs, in marshes or on weedy shores. The materials are sea-weed, tufts of grass or rushes, or the *débris* of the shore, heaped together with as little order and constructive art as might be expected from the webbed feet and clumsy bill of these birds, the latter better adapted for seizing fish than for forming a delicate nest. The long-legged, broad-billed flamingo, who is continually stalking over muddy flats in search of food, heaps up the mud into a conical stool, on the top of which it lays its eggs. The bird can thus sit upon them conveniently, and they are kept dry, out of reach of the tides.

Now I believe that throughout the whole class of birds the same general principles will be found to hold good, sometimes distinctly, sometimes more obscurely apparent, according as the habits of the species are more marked, or their structure more peculiar. It is true that, among birds differing but little in structure or habits, we see considerable diversity in the mode of nesting, but we are now so well assured that important changes of climate and of surface have occured within the period of existing species, that it is by no means difficult to see how such differences have arisen. Simple habits are known to be hereditary, and as the area now occupied by each species is different from that of every other, we may be sure that such changes would act differently upon each, and would often bring together species which had acquired their peculiar habits in distinct regions and under different conditions.

How do Young Birds learn to Build their First Nest?

But it is objected, birds do not *learn* to make their nest as man does to build, for all birds will make exactly the same nest as the rest of their species, even if they have never seen one, and it is instinct alone that can enable them to do this. No doubt this would be instinct if it were true, and I simply ask for proof of the fact. This point, although so important to the question at issue, is always assumed without proof, and even against proof, for what facts there are, are opposed to it. Birds brought up from the egg in cages do not make the characteristic nest of their species, even though the proper materials are supplied them, and often make no nest at all, but rudely heap together a quantity of materials; and the experiment has never been fairly tried, of turning out a pair of birds so brought up, into an enclosure covered with netting, and watching the result of their untaught attempts at nest-making. With regard to the songs of birds, however, which is thought to be equally instinctive, the experiment has been tried, and it is found that young birds never have the song peculiar to their species if they have not heard it, whereas they acquire very easily the song of almost any other bird with which they are associated.

Do Birds sing by Instinct or by Imitation?

The Hon. Daines Barrington was of opinion that "notes in birds are no more innate than language is in man, and depend entirely on the master under which they are bred, *as far as their organs will enable them to imitate* the sounds which they have frequent opportunities of hearing." He has given an account of his experiments in the "Philosophical Transactions" for 1773 (Vol. 63); he says: "I have educated nestling linnets under the three best singing larks—the skylark, woodlark, and titlark, every one of which, instead of the linnet's song, adhered entirely to that of their respective instructors. When the note of the titlark linnet was thoroughly fixed, I hung the bird in a room with two common linnets for a quarter of a year, which were full in song; the titlark linnet, however, did not borrow any passage from the linnet's

song, but adhered stedfastly to that of the titlark."
He then goes on to say that birds taken from the nest
at two or three weeks old have already learnt the call-
note of their species. To prevent this the birds must
be taken from the nest when a day or two old, and he
gives an account of a goldfinch which he saw at
Knighton in Radnorshire, and which sang exactly like
a wren, without any portion of the proper note of its
species. This bird had been taken from the nest at
two or three days old, and had been hung at a window
opposite a small garden, where it had undoubtedly
acquired the notes of the wren without having any
opportunity of learning even the call of the goldfinch.

He also saw a linnet, which had been taken from
the nest when only two or three days old, and which,
not having any other sounds to imitate, had learnt
almost to articulate, and could repeat the words
" Pretty Boy," and some other short sentences.

Another linnet was educated by himself under a
vengolina (a small African finch, which he says sings
better than any foreign bird but the American mock-
ing bird), and it imitated its African master so exactly
that it was impossible to distinguish the one from the
other.

Still more extraordinary was the case of a common
house sparrow, which only chirps in a wild state, but
which learnt the song of the linnet and goldfinch by
being brought up near those birds.

The Rev. W. H. Herbert made similar observations,
and states that the young whinchat and wheatear,

which have naturally little variety of song, are ready in confinement to learn from other species, and become much better songsters. The bullfinch, whose natural notes are weak, harsh, and insignificant, has nevertheless a wonderful musical faculty, since it can be taught to whistle complete tunes. The nightingale, on the other hand, whose natural song is so beautiful, is exceedingly apt in confinement to learn that of other birds instead. Bechstein gives an account of a redstart which had built under the eaves of his house, which imitated the song of a caged chaffinch in a window underneath, while another in his neighbour's garden repeated some of the notes of a blackcap, which had a nest close by.

These facts, and many others which might be quoted, render it certain that the peculiar notes of birds are acquired by imitation, as surely as a child learns English or French, not by instinct, but by hearing the language spoken by its parents.

It is especially worthy of remark that, for young birds to acquire a new song correctly, they must be taken out of hearing of their parents very soon, for in the first three or four days they have already acquired some knowledge of the parent notes, which they will afterwards imitate. This shows that very young birds can both hear and remember, and it would be very extraordinary if, after they could see, they could neither observe nor recollect, and could live for days and weeks in a nest and know nothing of its materials and the manner of its construction. During

the time they are learning to fly and return often to
the nest, they must be able to examine it inside and
out in every detail, and as we have seen that their
daily search for food invariably leads them among the
materials of which it is constructed, and among places
similar to that in which it is placed, is it so very
wonderful that when they want one themselves they
should make one like it? How else, in fact, should
they make it? Would it not be much more remarkable if they went out of their way to get materials
quite different from those used in the parent nest,
if they arranged them in a way they had seen no
example of, and formed the whole structure differently
from that in which they themselves were reared, and
which we may fairly presume is that which their whole
organization is best adapted to put together with cele-
rity and ease? It has, however, been objected that
observation, imitation, or memory, can have nothing
to do with a bird's architectural powers, because the
young birds, which in England are born in May or
June, will proceed in the following April or May to
build a nest as perfect and as beautiful as that in
which it was hatched, although it could never have
seen one built. But surely the young birds *before*
they left the nest had ample opportunities of observing
its *form*, its *size*, its *position*, the *materials* of which
it was constructed, and the manner in which those
materials were arranged. Memory would retain these
observations till the following spring, when the materials would come in their way during their daily

search for food, and it seems highly probable that the older birds would begin building first, and that those born the preceding summer would follow their example, learning from them how the foundations of the nest are laid and the materials put together.*

Again, we have no right to assume that young birds generally pair together. It seems probable that in each pair there is most frequently only one bird born the preceding summer, who would be guided, to some extent, by its partner.

My friend, Mr. Richard Spruce, the well-known traveller and botanist, thinks this is the case, and has kindly allowed me to publish the following observations, which he sent me after reading my book.

How young Birds may learn to build Nests.

" Among the Indians of Peru and Ecuador, many of whose customs are relics of the semi-civilisation that prevailed before the Spanish conquest, it is usual for the young men to marry old women, and the young women old men. A young man, they say, accustomed to be tended by his mother, would fare ill if

---

* It has been very pertinently remarked by a friend, that, if young birds did observe the nest they were reared in, they would consider it to be a natural production like the leaves and branches and matted twigs that surrounded it, and could not possibly conclude that their parents had constructed the one and not the other. This may be a valid objection, and, if so, we shall have to depend on the mode of instruction described in the succeeding paragraphs, but the question can only be finally decided by a careful set of experiments.

he had only an ignorant young girl to take care of him ; and the girl herself would be better off with a man of mature years, capable of supplying the place of a father to her.

" Something like this custom prevails among many animals. A stout old buck can generally fight his way to the doe of his choice, and indeed of as many does as he can manage ; but a young buck ' of his first horns,' must either content himself with celibacy, or with some dame well-stricken in years.

" Compare the nearly parallel case of the domestic cock and of many other birds. Then consider the consequences amongst birds that pair, if an old cock sorts with a young hen and an old hen with a young cock, as I think is certainly the case with blackbirds and others that are known to fight for the youngest and handsomest females. One of each pair being already an ' old bird,' will be competent to instruct its younger partner (not only in the futility of ' chaff,' but) in the selection of a site for a nest and how to build it ; then, how eggs are hatched and young birds reared.

" Such, in brief, is my idea of how a bird on its first espousals may be taught the Whole Duty of the married state."

On this difficult point I have sought for information from some of our best field ornithologists, but without success, as it is in most cases impossible to distinguish old from young birds after the first year. I am informed, however, that the males of blackbirds, sparrows, and many other kinds fight furiously, and the conqueror of course has the choice of a mate. Mr. Spruce's view is at least as probable as the contrary one (that young birds, *as a rule*, pair together), and it is to some extent supported by the celebrated American observer, Wilson, who strongly insists on the variety in the nests of birds of the same species, some being so much better finished than others; and he believes *that the less perfect nests are built by the younger, the more perfect by the older, birds.*

At all events, till the crucial experiment is made, and a pair of birds raised from the egg without ever seeing a nest are shown to be capable of making one exactly of the parental type, I do not think we are justified in calling in the aid of an unknown and mysterious faculty to do that which is so strictly analogous to the house-building of savage man.

Again, we always assume that because a nest appears to us delicately and artfully built, that it therefore requires much special knowledge and acquired skill (or their substitute, instinct) in the bird who builds it. We forget that it is formed twig by twig and fibre by fibre, rudely enough at first, but crevices and irregularities, which must seem huge gaps and chasms in the eyes of the little builders, are filled up by twigs and stalks pushed in by slender beak and active foot, and that the wool, feathers, or horsehair are laid thread by thread, so that the result seems a marvel of ingenuity to us, just as would the rudest Iinand hut to a native of Brobdignag.

Levaillant has given an account of the process of nest-building by a little African warbler, which sufficiently shows that a very beautiful structure may be produced with very little art. The foundation was laid of moss and flax interwoven with grass and tufts of cotton, and presented a rude mass, five or six inches in diameter, and four inches thick. This was pressed and trampled down repeatedly, so as at last to make it into a kind of felt. The birds pressed it with their bodies, turning round upon them in every direction, so as to get it quite firm and smooth before raising the sides. These were added bit by bit, trimmed and beaten with the wings and feet, so as to felt the whole together, projecting fibres being now and then worked in with the bill. By these simple and apparently inefficient means, the inner surface of the nest was rendered almost as smooth and compact as a piece of cloth.

Man's Works mainly Imitative.

But look at civilised man! it is said; look at Grecian, and Egyptian, and Roman, and Gothic, and modern Architecture! What advance! what improvement! what refinements! This is what reason leads to, whereas birds remain for ever stationary. If, however, such advances as these are required, to prove the effects of reason as contrasted with instinct, then all savage and many half-civilized tribes have no reason, but build instinctively quite as much as birds do.

Man ranges over the whole earth, and exists under the most varied conditions, leading necessarily to equally varied habits. He migrates—he makes wars and conquests—one race mingles with another—different customs are brought into contact—the habits of a migrating or conquering race are modified by the different circumstances of a new country. The civilized race which conquered Egypt must have developed its mode of building in a forest country where timber was abundant, for it is not probable, that the idea of cylindrical columns originated in a country destitute of trees. The pyramids might have been built by an indigenous race, but not the temples of El Uksor and Karnak. In Grecian architecture, almost every characteristic feature can be traced to an origin in wooden buildings. The columns, the architrave, the frieze, the fillets, the cantelevers, the form of the roof, all point to an origin in some southern forest-clad country, and strikingly corroborate the view derived from philology, that Greece was colonised from north-western India. But to erect columns and span them with huge blocks of stone, or marble, is not an act of reason, but one of pure unreasoning imitation. The arch is the only true and reasonable mode of covering over wide spaces with stone, and therefore, Grecian architecture, however exquisitely beautiful, is false in principle, and is by no means a good example of the application of reason to the art of building. And what do most of us do at the present day but imitate the buildings of those that have gone before us? We have not even been able to discover or develope any definite style of building best suited for us. We have no characteristic national style of architecture, and to that extent are even below the birds, who have each their characteristic form of nest, exactly adapted to their wants and habits.

Birds do Alter and Improve their Nests when altered Conditions require it.

The great uniformity in the architecture of each species of bird which has been supposed to prove a nest-building instinct, we may, therefore, fairly impute to the uniformity of the conditions under which each species lives. Their range is often very limited, and they very seldom permanently change their country, so as to be placed in new conditions. When, however, new conditions do occur, they take advantage of them just as freely and wisely as man could do. The chimney and house-swallows are a standing proof of a change of habit since chimneys and houses were built, and in America this change has taken place within about three hundred years. Thread and worsted are now used in many nests instead of wool and horsehair, and the jackdaw shows an affection for the church steeple which can hardly be explained by instinct. In the more thickly populated parts of the United States, the Baltimore oriole uses all sorts of pieces of string, skeins of silk, or the gardener's bass, to weave into its fine pensile nest, instead of the single hairs and vegetable fibres it has painfully to seek in wilder regions; and Wilson, a most careful observer, believes that it improves in nest-building by practice—the older birds making the best nests. The purple martin takes possession of empty gourds or small boxes, stuck up for its reception in almost every village and farm in America; and several of the American wrens will also build in cigar boxes, with a small hole cut in them, if placed in a suitable situation. The orchard oriole of the United States offers us an excellent example of a bird which modifies its nest according to circumstances. When built among firm and stiff branches the nest is very shallow, but if, as is often the case, it is suspended from the slender twigs of the weeping willow, it is made much deeper, so that when swayed about violently by the wind the young may not tumble out. It has been observed also, that the nests built in the warm Southern States are much slighter and more porous in texture than those in the colder regions of the north. Our own house-sparrow equally well adapts himself to circumstances. When he builds in trees, as he, no doubt, always did originally, he constructs a well-made domed nest, perfectly fitted to protect his young ones; but when he can find a convenient hole in a building or among thatch, or in any well-sheltered place, he takes much less trouble, and forms a very loosely-built nest.

A curious example of a recent change of habits has occurred in Jamaica. Previous to 1854, the palm swift (Tachornis phænicobea) inhabited exclusively the palm trees in a few districts in the island. A colony then established themselves in two cocoa-nut palms in Spanish Town, and remained there till 1857, when one tree was blown down, and the other stripped of its foliage. Instead of now seeking out other palm trees, the swifts drove out the swallows who built in the Piazza of the House of Assembly, and took possession of it, building their nests on the tops of the end walls and at the angles formed by the beams and joists, a place which they continue to occupy in considerable numbers. It is remarked that here they form their nest with much less elaboration than when built in the palms, probably from being less exposed.

A still more curious example of change and improvement in nest building was published by Mr. F. A. Pouchet, in the tenth number of the *Comptes Rendus* for 1870, just as the first edition of this work appeared. Forty years ago M. Pouchet had himself collected nests of the House-Martin or Window-Swallow (*Hirundo urbica*) from old buildings at Rouen, and deposited them in the museum of that city. On recently obtaining some more nests he was surprised, on comparing them with the old ones, to find that they exhibited a decided change of form and structure. This led him to investigate the matter more closely. The changed nests had been obtained from houses in a newly erected quarter of the city, and he found that all the nests in the newly-built streets were of the new form. But on visiting the churches and older buildings, and some rocks where these birds build, he found many nests of the old type along with some of the new pattern. He then examined all the figures and descriptions of the older naturalists, and found that they invariably represented the older form only.

The difference between the two forms he states to be as follows. In the old form the nest is a portion of a globe—when situated in the upper angle of a window one-fourth of a hemisphere—and the opening is very small and circular, being of a size just sufficient to allow the body of the bird to pass. In the new form the nest is much wider in proportion to its height, being a segment of a depressed spheroid, and the aperture is very wide and shallow, and close to the horizontal surface to which the nest is attached above.

M. Pouchet thinks that the new form is an undoubted improvement on the old. The nest has a wider bottom and must allow the young ones to have more freedom of motion than in the old narrower, and deeper nests, and its wide aperture allows the young birds to peep out and breathe the fresh air. This is so wide as to serve as a sort of balcony for them, and two young ones can often be seen on it without interfering with the passage in and out of the old birds. At the same time, by being so close to the roof, it is a better protection against rain, against cold, and against enemies, than the small round hole of the old nests. Here, then, we have an improvement in nest building, as well marked as any improvement that takes place in human dwellings in so short a time.

But perfection of structure and adaptation to purpose, are not universal characteristics of birds' nests, since there are decided imperfections in the nesting of many birds which are quite compatible with our present theory, but are hardly so with that of instinct, which is supposed to be infallible. The Passenger pigeon of America often crowds the branches with its nests till they break, and the ground is strewn with shattered nests, eggs, and young birds. Rooks' nests are often so imperfect that during high winds the eggs fall out; but the Window-Swallow is the most unfortunate in this respect, for White, of Selborne, informs us that he has seen them build, year after year, in places where their nests are liable to be washed away by a heavy rain and their young ones destroyed.

Conclusion.

A fair consideration of all these facts will, I think, fully support the statement with which I commenced, and show, that the mental faculties exhibited by birds in the construction of their nests, are the same in kind as those manifested by mankind in the formation of their dwellings. These are, essentially, imitation, and a slow and partial adaptation to new conditions. To compare the work of birds with the highest manifestations of human art and science, is totally beside the question. I do not maintain that birds are gifted with reasoning faculties at all approaching in variety and extent to those of man. I simply hold that the phenomena presented by their mode of building their nests, when fairly compared with those exhibited by the great mass of mankind in building their houses, indicate no essential difference in the kind or nature of the mental faculties employed. If instinct means anything, it means the capacity to perform some complex act without teaching or experience. It implies innate ideas of a very definite kind, and, if established, would overthrow Mr. Mill's sensationalism and all the modern philosophy of experience. That the existence of true instinct may be established in other cases is not impossible, but in the particular instance of birds' nests, which is usually considered one of its strongholds, I cannot find a particle of evidence to show the existence of anything beyond those lower reasoning and imitative powers, which animals are universally admitted to possess.

VII.

A THEORY OF BIRDS' NESTS;
SHOWING THE RELATION OF CERTAIN DIFFERENCES OF
COLOUR IN FEMALE BIRDS, TO THEIR MODE OF
NIDIFICATION.

THE habit of forming a more or less elaborate structure for the reception of their eggs and young, must undoubtedly be looked upon as one of the most remarkable and interesting characteristics of the class of birds. In other classes of vertebrate animals, such structures are few and exceptional, and never attain to the same degree of completeness and beauty. Birds' nests have, accordingly, attracted much attention, and have furnished one of the stock arguments to prove the existence of a blind but unerring instinct in the lower animals. The very general belief that every bird is enabled to build its nest, not by the ordinary faculties of observation, memory, and imitation, but by means of some innate and mysterious impulse, has had the bad effect of withdrawing attention from the very evident relation that exists between the structure, habits, and intelligence of birds, and the kind of nests they construct.

In the preceding essay I have detailed several of these relations, and they teach us, that a consideration of the structure, the food, and other specialities of a bird's existence, will give a clue, and sometimes a very complete one, to the reason why it builds its nest of certain materials, in a definite situation, and in a more or less elaborate manner.

I now propose to consider the question from a more general point of view, and to discuss its application to some important problems in the natural history of birds.

Changed Conditions and persistent Habits as influencing Nidification.

Besides the causes above alluded to, there are two other factors whose effect in any particular case we can only vaguely guess at, but which must have had an important influence in determining the existing details of nidification. These are—changed conditions of existence, whether internal or external, and the influence of hereditary or imitative habit; the first inducing alterations in accordance with changes of organic structure, of climate, or of the surrounding fauna and flora; the other preserving the peculiarities so produced, even when changed conditions render them no longer necessary. Many facts have been already given which show that birds do adapt their nests to the situations in which they place them, and the adoption of eaves, chimneys, and boxes, by swallows, wrens, and many other birds, shows that they are always ready to take advantage of changed conditions. It is probable, therefore, that a permanent change of climate would cause many birds to modify the form or materials of their abodes, so as better to protect their young. The introduction of new enemies to eggs or young birds, might introduce many alterations tending to their better concealment. A change in the vegetation of a country, would often necessitate the use of new materials. So, also, we may be sure, that as a species slowly became modified in any external or internal characters, it would necessarily change in some degree its mode of building. This effect would be produced by modifications of the most varied nature; such as the power and rapidity of flight, which must often determine the distance to which a bird will go to obtain materials for its nest; the capacity of sustaining itself almost motionless in the air, which must sometimes determine the position in which a nest can be built; the strength and grasping power of the foot in relation to the weight of the bird, a power absolutely essential to the constructor of a delicately-woven and well-finished nest; the length and fineness of the beak, which has to be used like a needle in building the best textile nests; the length and mobility of the neck, which is needful for the same purpose; the possession of a salivary secretion like that used in the nests of many of the swifts and swallows, as well as that of the song-thrush—peculiarities of habits, which ultimately depend on structure, and which often determine the material most frequently met with or most easily to be obtained. Modifications in any of these characters would necessarily lead, either to a change in the materials of the nest, or in the mode of combining them in the finished structure, or in the form or position of that structure.

During all these changes, however, certain specialities of nest-building would continue, for a shorter or a longer time after the causes which had necessitated them had passed away. Such records of a vanished past meet us everywhere, even in man's works, notwithstanding his boasted reason. Not only are the main features of Greek architecture, mere reproductions in stone of what were originally parts of a wooden building, but our modern copyists of Gothic architecture often build solid buttresses capped with weighty pinnacles, to support a wooden roof which has no outward thrust to render them necessary; and even think they ornament their buildings by adding sham spouts of carved stone, while modern waterpipes, stuck on without any attempt at harmony, do the real duty. So, when railways superseded coaches, it was thought necessary to build the first-class carriages to imitate a number of coach-bodies joined together; and the arm-loops for each passenger to hold on by, which were useful when bad roads made every journey a succession of jolts and lurches, were continued on our smooth macadamised mail-routes, and, still more absurdly, remain to this day in our railway carriages, the relic of a kind of locomotion we can now hardly realize. Another good example is to be seen in our boots. When elastic sides came into fashion we had been so long used to fasten them with buttons or laces, that a boot without either looked bare and unfinished,

and accordingly the makers often put on a row of
useless buttons or imitation laces, because habit rendered the appearance of them necessary to us. It is
universally admitted that the habits of children and of
savages give us the best clue to the habits and mode
of thought of animals; and every one must have
observed how children at first imitate the actions of
their elders, without any regard to the use or applicability of the particular acts. So, in savages, many
customs peculiar to each tribe are handed down from
father to son merely by the force of habit, and are
continued long after the purpose which they originally served has ceased to exist. With these and a
hundred similar facts everywhere around us, we may
fairly impute much of what we cannot understand in
the details of Bird-Architecture to an analogous cause.
If we do not do so, we must assume, either that birds
are guided in every action by pure reason to a far
greater extent than men are, or that an infallible instinct leads them to the same result by a different
road. The first theory has never, that I am aware
of, been maintained by any author, and I have already
shown that the second, although constantly assumed,
has never been proved, and that a large body of facts
is entirely opposed to it. One of my critics has, indeed, maintained that I admit "instinct" under the
term "hereditary habit;" but the whole course of my
argument shows that I do not do so. Hereditary
habit is, indeed, the same as instinct when the term
is applied to some simple action dependent upon a

peculiarity of structure which is hereditary; as when the descendants of tumbler pigeons tumble, and the descendants of pouter pigeons pout. In the present case, however, I compare it strictly to the hereditary, or more properly, persistent or imitative, habits of savages, in building their houses as their fathers did. Imitation is a lower faculty than invention. Children and savages imitate before they originate; birds, as well as all other animals, do the same.

The preceding observations are intended to show, that the exact mode of nidification of each species of bird is probably the result of a variety of causes, which have been continually inducing changes in accordance with changed organic or physical conditions. The most important of these causes seem to be, in the first place, the structure of the species, and, in the second, its environment or conditions of existence. Now we know, that every one of the characters or conditions included under these two heads is variable. We have seen that, on the large scale, the main features of the nest built by each group of birds, bears a relation to the organic structure of that group, and we have, therefore, a right to infer, that as structure varies, the nest will vary also in some particular corresponding to the changes of structure. We have seen also, that birds change the position, the form, and the construction of their nest, whenever the available materials or the available situations, vary naturally or have been altered by man; and we have, therefore, a right to infer that similar changes have taken place, when, by a natural process, external conditions have become in any way permanently altered. We must remember, however, that all these factors are very stable during many generations, and only change at a rate commensurate with those of the great physical features of the earth as revealed to us by geology ; and we may, therefore, infer that the form and construction of nests, which we have shown to be dependent on them, are equally stable. If, therefore, we find less important and more easily modified characters than these, so correlated with peculiarities of nidification as to indicate that one is probably the cause of the other, we shall be justified in concluding that these variable characters are dependent on the mode of nidification, and not that the form of the nest has been determined by these variable characters. Such a correlation I am now about to point out.

Classification of Nests.

For the purpose of this inquiry it is necessary to group nests into two great classes, without any regard to their most obvious differences or resemblances, but solely looking to the fact of whether the contents (eggs, young, or sitting bird) are hidden or exposed to view. In the first class we place all those in which the eggs and young are completely hidden, no matter whether this is effected by an elaborate covered structure, or by depositing the eggs in some hollow tree or burrow underground. In the second, we group all in which the eggs, young, and sitting bird are exposed to view, no matter whether there is the most beautifully formed nest, or none at all.   Kingfishers, which build almost invariably in holes in banks; Woodpeckers and Parrots, which build in hollow trees; the Icteridæ of America, which all make beautiful covered and suspended nests; and our own Wren, which builds a domed nest, are examples of the former; while our Thrushes, Warblers, and Finches, as well as the Crowshrikes, Chatterers, and Tanagers of the tropics, together with all Raptorial birds and Pigeons, and a vast number of others in every part of the world, all adopt the latter mode of building.

It will be seen that this division of birds according to their nidification, bears little relation to the character of the nest itself.   It is a functional not a structural classification.   The most rude and the most perfect specimens of bird-architecture are to be found in both sections.   It has, however, a certain relation to natural affinities, for large groups of birds, undoubtedly allied, fall into one or the other division exclusively.   The species of a genus or of a family are rarely divided between the two primary classes, although they are frequently divided between the two very distinct modes of nidification that exist in the first of them.

All the Scansorial or climbing, and most of the Fissirostral or wide-gaped birds, for example, build concealed nests; and, in the latter group, the two families which build open nests, the Swifts and the Goat-suckers, are undoubtedly very widely separated from the other families with which they are associated in our classifications. The Tits vary much in their mode of nesting, some making open nests concealed in a hole, while others build domed or even pendulous covered nests, but they all come under the same class. Starlings vary in a similar way. The talking Mynahs, like our own starlings, build in holes, the glossy starlings of the East (of the genus Calornis) form a hanging covered nest, while the genus Sturnopastor builds in a hollow tree. One of the most striking cases in which one family of birds is divided between the two classes, is that of the Finches; for while most of the European species build exposed nests, many of the Australian finches make them dome-shaped.

Sexual differences of Colour in Birds.

Turning now from the nests to the creatures who make them, let us consider birds themselves from a somewhat unusual point of view, and form them into separate groups, according as both sexes, or the males only, are adorned with conspicuous colours.

The sexual differences of colour and plumage in birds are very remarkable, and have attracted much attention; and, in the case of polygamous birds, have been well explained by Mr. Darwin's principle of sexual selection. We can, to a great extent, understand how male Pheasants and Grouse have acquired their more brilliant plumage and greater size, by the continual rivalry of the males both in strength and beauty; but this theory does not throw any light on the causes which have made the female Toucan, Bee-eater, Parroquet, Macaw and Tit, in almost every case as gay and brilliant as the male, while the gorgeous Chatterers, Manakins, Tanagers, and Birds of Paradise, as well as our own Blackbird, have mates so dull and inconspicuous that they can hardly be recognised as belonging to the same species.

The Law which connects the Colours of Female Birds with the mode of Nidification.

The above-stated anomaly can, however, now be explained by the influence of the mode of nidification, since I find that, with but very few exceptions, it is the rule—*that when both sexes are of strikingly gay and conspicuous colours, the nest is of the first class, or such as to conceal the sitting bird; while, whenever there is a striking contrast of colours, the male being gay and conspicuous, the female dull and obscure, the nest is open and the sitting bird exposed to view.* I will now proceed to indicate the chief facts that support this statement, and will afterwards explain the manner in which I conceive the relation has been brought about.

We will first consider those groups of birds in which the female is gaily or at least conspicuously coloured, and is in most cases exactly like the male.

1. Kingfishers (Alcedinidæ). In some of the most brilliant species of this family the female exactly resembles the male; in others there is a sexual difference, but it rarely tends to make the female less conspicuous. In some, the female has a band across the breast, which is wanting in the male, as in the beautiful Halcyon

diops of Ternate. In others the band is rufous in the female, as in several of the American species; while in Dacelo gaudichaudii, and others of the same genus, the tail of the female is rufous, while that of the male is blue. In most kingfishers the nest is in a deep hole in the ground; in Tanysiptera it is said to be in a hole in the nests of termites, or sometimes in crevices under overhanging rocks.

2. Motmots (Momotidæ). In these showy birds the sexes are exactly alike, and the nest in a hole under ground.

3. Puff-birds (Bucconidæ). These birds are often gaily coloured; some have coral-red bills; the sexes are exactly alike, and the nest is in a hole in sloping ground.

4. Trogons (Trogonidæ). In these magnificent birds the females are generally less brightly coloured than the males, but are yet often gay and conspicuous. The nest is in a hole of a tree.

5. Hoopoes (Upupidæ). The barred plumage and long crests of these birds render them conspicuous. The sexes are exactly alike, and the nest is in a hollow tree.

6. Hornbills (Bucerotidæ). These large birds have enormous coloured bills, which are generally quite as well coloured and conspicuous in the females. Their nests are always in hollow trees, where the female is entirely concealed.

7. Barbets (Capitonidæ). These birds are all very gaily-coloured, and, what is remarkable, the most brilliant patches of colour are disposed about the head and

R

neck, and are very conspicuous. The sexes are exactly alike, and the nest is in a hole of a tree.

8. Toucans (Rhamphastidæ). These fine birds are coloured in the most conspicuous parts of their body, especially on the large bill, and on the upper and lower tail coverts, which are crimson, white, or yellow. The sexes are exactly alike, and they always build in a hollow tree.

9. Plantain-eaters (Musophagidæ). Here again the head and bill are most brilliantly coloured in both sexes, and the nest is in a hole of a tree.

10. Ground cuckoos (Centropus). These birds are often of conspicuous colours, and are alike in both sexes. They build a domed nest.

11. Woodpeckers (Picidæ). In this family the females often differ from the males, in having a yellow or white, instead of a crimson crest, but are almost as conspicuous. They all nest in holes in trees.

12. Parrots (Psittaci). In this great tribe, adorned with the most brilliant and varied colours, the rule is, that the sexes are precisely alike, and this is the case in the most gorgeous families, the lories, the cockatoos, and the macaws; but in some there is a sexual difference of colour to a slight extent. All build in holes, mostly in trees, but sometimes in the ground, or in white ants' nests. In the single case in which the nest is exposed, that of the Australian ground parrot, Pezoporus formosus, the bird has lost the gay colouring of its allies, and is clothed in sombre and completely protective tints of dusky green and black.

13. Gapers (Eurylæmidæ). In these beautiful Eastern birds, somewhat allied to the American chatterers, the sexes are exactly alike, and are adorned with the most gay and conspicuous markings. The nest is a woven structure, *covered over*, and suspended from the extremities of branches over water.

14. Pardalotus (Ampelidæ). In these Australian birds the females differ from the males, but are often very conspicuous, having brightly-spotted heads. Their nests are sometimes dome-shaped, sometimes in holes of trees, or in burrows in the ground.

15. Tits (Paridæ). These little birds are always pretty, and many (especially among the Indian species) are very conspicuous. They always have the sexes alike, a circumstance very unusual among the smaller gaily-coloured birds of our own country. The nest is always covered over or concealed in a hole.

16. Nuthatches (Sitta). Often very pretty birds, the sexes alike, and the nest in a hole.

17. —— (Sittella). The female of these Australian nuthatches is often the most conspicuous, being white- and black-marked. The nest is, according to Gould, " completely concealed among upright twigs connected together."

18. Creepers (Climacteris). In these Australian creepers the sexes are alike, or the female most conspicuous, and the nest is in a hole of a tree.

19. Estrelda, Amadina. In these genera of Eastern and Australian finches the females, although more or less different from the males, are still very conspicuous

having a red rump, or being white spotted. They differ from most others of the family in building domed nests.

20. Certhiola. In these pretty little American creepers the sexes are alike, and they build a domed nest.

21. Mynahs (Sturnidæ). These showy Eastern starlings have the sexes exactly alike. They build in holes of trees.

22. Calornis (Sturnidæ). These brilliant metallic starlings have no sexual differences. They build a pensile covered nest.

23. Hangnests (Icteridæ). The red or yellow and black plumage of most of these birds is very conspicuous, and is exactly alike in both sexes. They are celebrated for their fine purse-shaped pensile nests.

It will be seen that this list comprehends six important families of Fissirostres, four of Scansores, the Psittaci, and several genera, with three entire families of Passeres, comprising about twelve hundred species, or about one-seventh of all known birds.

The cases in which, whenever the male is gaily coloured, the female is much less gay or quite inconspicuous, are exceedingly numerous, comprising, in fact, almost all the bright-coloured Passeres, except those enumerated in the preceding class. The following are the most remarkable :—

1. Chatterers (Cotingidæ). These comprise some of the most gorgeous birds in the world, vivid blues,

rich purples, and bright reds, being the most characteristic colours. The females are always obscurely tinted, and are often of a greenish hue, not easily visible among the foliage.

2. Manakins (Pipridæ). These elegant birds, whose caps or crests are of the most brilliant colours, are usually of a sombre green in the female sex.

3. Tanagers (Tanagridæ). These rival the chatterers in the brilliancy of their colours, and are even more varied. The females are generally of plain and sombre hues, and always less conspicuous than the males.

In the extensive families of the warblers (Sylviadæ), thrushes (Turdidæ), flycatchers (Muscicapidæ), and shrikes (Laniadæ), a considerable proportion of the species are beautifully marked with gay and conspicuous tints, as is also the case in the Pheasants and Grouse ; but in every case the females are less gay, and are most frequently of the very plainest and least conspicuous hues. Now, throughout *the whole of these families the nest is open,* and I am not aware of a single instance in which any one of these birds builds a *domed nest,* or places it in a *hole of a tree,* or *underground,* or in any place where it is effectually concealed.

In considering the question we are now investigating, it is not necessary to take into account the larger and more powerful birds, because these seldom depend much on concealment to secure their safety. In the raptorial birds bright colours are as a rule absent ; and their structure and habits are such as not to require any special protection for the female. The larger waders are sometimes very brightly coloured in both sexes; but they are probably little subject to the attacks of enemies, since the scarlet ibis, the most conspicuous of birds, exists in immense quantities in South America. In game birds and water-fowl, however, the females are often very plainly coloured, when the males are adorned with brilliant hues; and the abnormal family of the Megapodidæ offers us the interesting fact of an identity in the colours of the sexes (which in Mega-cephalon and Talegalla are somewhat conspicuous), in conjunction with the habit of not sitting on the eggs at all.

What the Facts Teach us.

Taking the whole body of evidence here brought forward, embracing as it does almost every group of bright-coloured birds, it will, I think, be admitted that the relation between the two series of facts in the colouring and nidification of birds has been sufficiently established. There are, it is true, a few apparent and some real exceptions, which I shall consider presently; but they are too few and unimportant to weigh much against the mass of evidence on the other side, and may for the present be neglected. Let us then consider what we are to do with this unexpected set of correspondences between groups of phenomena which, at first sight, appear so disconnected. Do they fall in with any other groups of natural phenomena? Do they teach us anything of the way in which nature works, and give us any insight into the causes which have brought about the marvellous variety, and beauty, and harmony of living things? I believe we can answer these questions in the affirmative; and I may mention, as a sufficient proof that these are not isolated facts, that I was first led to see their relation to each other by the study of an analogous though distinct set of phenomena among insects, that of protective resemblance and "mimicry."

On considering this remarkable series of corresponding facts, the first thing we are taught by them seems to be, that there is no incapacity in the female sex among birds, to receive the same bright hues and strongly contrasted tints with which their partners are so often decorated, since whenever they are *protected and concealed* during the period of incubation *they are similarly adorned.* The fair inference is, that it is chiefly due to the absence of protection or concealment during this important epoch, that gay and conspicuous tints are withheld or left undeveloped. The mode in which this has been effected is very intelligible, if we admit the action of natural and sexual selection. It would appear from the numerous cases in which both sexes are adorned with equally brilliant colours (while both sexes are rarely armed with equally developed offensive and defensive weapons when not required for individual safety), that the normal action of "sexual selection" is to develop colour and beauty in both sexes, by the preservation and multiplication of all varieties of colour in either sex which are pleasing to the other. Several very close observers of the habits of animals have assured me, that male birds and quadrupeds do often take very strong likes and dislikes to individual females, and we can hardly believe that the one sex (the female) can have a general taste for colour while the other has no such taste. However this may be, the fact remains, that in a vast number of cases the female acquires as brilliant and as varied colours as the male, and therefore most probably acquires them in the same way as the male does; that is, either because the colour is useful to it, or is correlated with some useful variation, or is pleasing to the other sex. The only remaining supposition is that it is transmitted from the other sex, without being of any use. From the number of examples above adduced of bright colours in the female, this would imply that colour-characters acquired by one sex are generally (but not necessarily) transmitted to the other. If this be the case it will, I think, enable us to explain the phenomena, even if we do not admit that the male bird is ever influenced in the choice of a mate by her more gay or perfect plumage.

The female bird, while sitting on her eggs in an uncovered nest, is much exposed to the attacks of enemies, and any modification of colour which rendered her more conspicuous would often lead to her destruction and that of her offspring. All variations of colour in this direction in the female, would therefore sooner or later be eliminated, while such modifications as rendered her inconspicuous, by assimilating her to surrounding objects, as the earth or the foliage, would, on the whole, survive the longest, and thus lead to the attainment of those brown or green and inconspicuous tints, which form the colouring (of the upper surface at least), of the vast majority of female birds which sit upon open nests.

This does not imply, as some have thought, that all female birds were once as brilliant as the males. The change has been a very gradual one, generally dating from the origin of genera or of larger groups, but there can be no doubt that the remote ancestry of birds having great sexual differences of colour, were nearly or quite alike, sometimes (perhaps in most cases) more nearly resembling the female, but occasionally perhaps being nearer what the male is now. The young birds (which usually resemble the females) will probably give some idea of this ancestral type, and it is well known that the young of allied species and of different sexes are often undistinguishable.

Colour more variable than Structure or Habits, and therefore the Character which has generally been Modified.

At the commencement of this essay, I have endeavoured to prove, that the characteristic differences and the essential features of birds' nests, are dependent on the structure of the species and upon the present and past conditions of their existence. Both these factors are more important and less variable than colour; and we must therefore conclude that in most cases the mode of nidification (dependent on structure and environment) has been the cause, and not the effect, of the similarity or differences of the sexes as regards colour. When the confirmed habit of a group of birds, was to build their nests in holes of trees like the toucans, or in holes in the ground like the kingfishers, the protection the female thus obtained, during the important and dangerous time of incubation, placed the two sexes on an equality as regards exposure to attack, and allowed "sexual selection," or any other cause, to act unchecked in the development of gay colours and conspicuous markings in both sexes.

When, on the other hand (as in the Tanagers and Flycatchers), the habit of the whole group was to build open cup-shaped nests in more or less exposed situations, the production of colour and marking in the female, by whatever cause, was continually checked by its rendering her too conspicuous, while in the male it had free play, and developed in him the most gorgeous hues. This, however, was not perhaps universally the case; for where there was more than usual intelligence and capacity for change of habits, the danger the female was exposed to by a partial brightness of colour or marking might lead to the construction of a concealed or covered nest, as in the case of the Tits and Hang-nests. When this occurred, a special protection to the female would be no longer necessary; so that the acquisition of colour and the modification of the nest, might in some cases act and react on each other and attain their full development together.

*Exceptional Cases confirmatory of the above
Explanation.*

There exist a few very curious and anomalous facts
in the natural history of birds, which fortunately serve
as crucial tests of the truth of this mode of explaining
the inequalities of sexual colouration. It has been long
known, that in some species the males either assisted in,
or wholly performed, the act of incubation. It has also
been often noticed, that in certain birds the usual sexual
differences were reversed, the male being the more
plainly coloured, the female more gay and often larger.
I am not, however, aware that these two anomalies had
ever been supposed to stand to each other in the relation of cause and effect, till I adduced them in support
of my views of the general theory of protective adaptation. Yet it is undoubtedly the fact, that in the best
known cases in which the female bird is more conspicuously coloured than the male, it is either positively
ascertained that the latter performs the duties of incubation, or there are good reasons for believing such
to be the case. The most satisfactory example is that
of the Gray Phalarope (Phalaropus fulicarius), the
sexes of which are alike in winter, while in summer
the female instead of the male takes on a gay and
conspicuous nuptial plumage ; but the male performs
the duties of incubation, sitting upon the eggs, which
are laid upon the bare ground.

In the Dotterell (Eudromias morinellus) the female
is larger and more brightly coloured than the male ; and

here, also, it is almost certain that the latter sits upon the eggs. The Turnices of India also, have the female larger and often more brightly coloured; and Mr. Jerdon states, in his " Birds of India," that the natives report, that, during the breeding season, the females desert their eggs and associate in flocks, while the males are employed in hatching the eggs. In the few other cases in which the females are more brightly coloured, the habits are not accurately known. The case of the Ostriches and Emeus will occur to many as a difficulty, for here the male incubates, but is not less conspicuous than the female; but there are two reasons why the case does not apply;—the birds are too large to derive any safety from concealment, from enemies which would devour the eggs they can defend themselves by force, while to escape from their personal foes they trust to speed.

We find, therefore, that a very large mass of facts relating to the sexual colouration and the mode of nidification of birds, including some of the most extraordinary anomalies to be found in their natural history, can be shown to have an interdependent relation to each other, on the simple principle of the need of greater protection to that parent which performs the duties of incubation. Considering the very imperfect knowledge we possess of the habits of most extra-European birds, the exceptions to the prevalent rule are few, and generally occur in isolated species or in small groups; while several apparent exceptions can be shown to be really confirmations of the law.

Real or apparent Exceptions to the Law stated at
page 240.

The only marked exceptions I have been able to discover are the following :—

1. King crows (Dicrourus). These birds are of a glossy black colour with long forked tails. The sexes present no difference, and they build open nests. This apparent exception may probably be accounted for by the fact that these birds do not need the protection of a less conspicuous colour. They are very pugnacious, and often attack and drive away crows, hawks, and kites ; and as they are semi-gregarious in their habits, the females are not likely to be attacked while incubating.

2. Orioles (Oriolidæ). The true orioles are very gay birds ; the sexes are, in many Eastern species, either nearly or quite alike, and the nests are open. This is one of the most serious exceptions, but it is one that to some extent proves the rule ; for in this case it has been noticed, that the parent birds display excessive care and solicitude in concealing the nest among thick foliage, and in protecting their offspring by incessant and anxious watching. This indicates that the want of protection consequent on the bright colour of the female makes itself felt, and is obviated by an increased development of the mental faculties.

3. Ground thrushes (Pittidæ). These elegant and brilliantly-coloured birds are generally alike in both sexes, and build an open nest. It is curious, however,

that this is only an apparent exception, for almost all the bright colours are on the under surface, the back being usually olive green or brown, and the head black, with brown or whitish stripes, all which colours would harmonize with the foliage, sticks, and roots which surround the nest, built on or near the ground, and thus serve as a protection to the female bird.

4. Grallina Australis. This Australian bird is of strongly contrasted black and white colours. The sexes are exactly alike, and it builds an open clay nest in an exposed situation on a tree. This appears to be a most striking exception, but I am by no means sure that it is so. We require to know what tree it usually builds on, the colour of the bark or of the lichens that grow upon it, the tints of the ground, or of other surrounding objects, before we can say that the bird, when sitting on its nest, is really conspicuous. It has been remarked that small patches of white and black blend at a short distance to form grey, one of the commonest tints of natural objects.

5. Sunbirds (Nectarineidæ). In these beautiful little birds the males only are adorned with brilliant colours, the females being quite plain, yet they build covered nests in all the cases in which the nidification is known. This is a negative rather than a positive exception to the rule, since there may be other causes besides the need for protection, which prevent the female acquiring the gay colours of her mate, and there is one curious circumstance which tends to elucidate it. The male of Leptocoma zeylanica is said

to assist in incubation. It is possible, therefore, that the group may originally have used open nests, and some change of conditions, leading the male bird to sit, may have been followed by the adoption of a domed nest. This is, however, the most serious exception I have yet found to the general rule.

6. Superb warblers (Maluridæ). The males of these little birds are adorned with the most gorgeous colours, while the females are very plain, yet they make domed nests. It is to be observed, however, that the male plumage is nuptial merely, and is retained for a very short time; the rest of the year both sexes are plain alike. It is probable, therefore, that the domed nest is for the protection of these delicate little birds against the rain, and that there is some unknown cause which has led to the development of colour in the males only.

There is one other case which at first sight looks like an exception, but which is far from being one in reality, and deserves to be mentioned. In the beautiful Waxwing, (Bombycilla garrula,) the sexes are very nearly alike, and the elegant red wax tips to the wing-feathers are nearly, and sometimes quite, as conspicuous in the female as in the male. Yet it builds an open nest, and a person looking at the bird would say it ought according to my theory to cover its nest. But it is, in reality, as completely protected by its colouration as the most plainly coloured bird that flies. It breeds only in very high latitudes, and the nest, placed in fir-trees, is formed chiefly of lichens. Now the delicate gray and ashy and purplish hues of the head and back, together with the yellow of the wings and tail, are tints that exactly harmonize with the colours of various species of lichens, while the brilliant red wax tips exactly represent the crimson fructification of the common lichen, Cladonia cocci-fera. When sitting on its nest, therefore, the female bird will exhibit no colours that are not common to the materials of which it is constructed; and the several tints are distributed in about the same pro-portions as they occur in nature. At a short distance the bird would be indistinguishable from the nest it is sitting on, or from a natural clump of lichens, and will thus be completely protected.

I think I have now noticed all exceptions of any importance to the law of dependence of sexual colour on nidification. It will be seen that they are very few in number, compared with those which support the generalization; and in several cases there are circumstances in the habits or structure of the species that sufficiently explain them. It is remarkable also that I have found scarcely any *positive* exceptions, that is, cases of very brilliant or conspicuous female birds in which the nest was not concealed. Much less can there be shown any group of birds, in which the females are all of decidedly conspicuous colours on the upper surface, and yet sit in open nests. The many cases in which birds of dull colours in both sexes make domed or concealed nests, do not, of course, affect this theory one way or the other; since its purpose is only to account for the fact, that brilliant females of brilliant males are *always* found to have covered or hidden nests, while obscure females of brilliant males *almost always* have open and exposed nests. The fact that all classes of nests occur with dull coloured birds in both sexes merely shows, as I have strongly maintained, that in most cases the character of the nest determines the colouration of the female, and not *vice versâ.*

If the views here advocated are correct, as to the various influences that have determined the specialities of every bird's nest, and the general colouration of female birds, with their action and reaction on each other, we can hardly expect to find evidence more complete than that here set forth. Nature is such a tangled web of complex relations, that a series of correspondences running through hundreds of species, genera, and families, in every part of the system, can hardly fail to indicate a true casual connexion; and when, of the two factors in the problem, one can be shown to be dependent on the most deeply seated and the most stable facts of structure and conditions of life, while the other is a character universally admitted to be superficial and easily modified, there can be little doubt as to which is cause and which effect.

Various modes of Protection of Animals

But the explanation of the phenomenon here attempted does not rest alone on the facts I have been able now to adduce. In the essay on " Mimicry," it is shown how important a part the necessity for

s

protection has played, in determining the external form and colouration, and sometimes even the internal structure of animals.

As illustrating this latter point, I may refer to the remarkable hooked, branched, or star-like spiculæ in many sponges, which are believed to have the function chiefly, of rendering them unpalatable to other creatures. The Holothuridæ or sea-cucumbers possess a similar protection, many of them having anchor-shaped spicules embedded in their skin, as the Synapta; while others (Cuviera squamata) are covered with a hard calcareous pavement. Many of these are of a bright red or purple colour, and are very conspicuous, while the allied Trepang, or Beche-de-mer (Holothuria edulis), which is not armed with any such defensive weapons, is of a dull sand- or mud-colour, so as hardly to be distinguished from the sea bed on which it reposes. Many of the smaller marine animals are protected by their almost invisible transparency, while those that are most brightly coloured will be often found to have a special protection, either in stinging tentacles like Physalia, or in a hard calcareous crust, as in the star fishes.

Females of some Groups require and obtain more Protection than the Males.

In the struggle for existence incessantly going on, protection or concealment is one of the most general and most effectual means of maintaining life; and it is by modifications of colour that this protection can be

most readily obtained, since no other character is subject to such numerous and rapid variations. The case I
have now endeavoured to illustrate is exactly analogous
to what occurs among butterflies. As a general rule,
the female butterfly is of dull and inconspicuous colours,
even when the male is most gorgeously arrayed; but
when the species is protected from attack by a disagreeable odour, as in the Heliconidæ, Danaidæ and
Acrœidæ, both sexes display the same or equally brilliant hues. Among the species which gain a protection by imitating these, the very weak and slow-flying
Leptalides resemble them in both sexes, because both
sexes alike require protection, while in the more active
and strong-winged genera—Papilio, Pieris, and Diadema—it is generally the females only that mimic the
protected groups, and in doing so often become actually
more gay and more conspicuous than the males, thus
reversing the usual and in fact almost universal characters of the sexes. So, in the wonderful Eastern leaf-
insects of the genus Phyllium, it is the female only
that so marvellously imitates a green leaf; and in all
these cases the difference can be traced to the greater
need of protection for the female, on whose continued
existence, while depositing her eggs, the safety of the
race depends. In Mammalia and in reptiles, however
brilliant the colours may be, there is rarely any difference between that of the sexes, because the female is
not necessarily more exposed to attack than the male.
It may, I think, be looked upon as a confirmation of
this view, that no single case is known either in the

above-named genera—Papilio, Pieris, and Diadema—or in any other butterfly, of a male *alone*, mimicking one of the Danaidæ or Heliconidæ. Yet the necessary colour is far more abundant in the males, and variations always seem ready for any useful purpose. This seems to depend on the general law, that each species and each sex can only be modified just as far as is absolutely necessary for it to maintain itself in the struggle for existence, not a step further. A male insect by its structure and habits is less exposed to danger, and also requires less protection than the female. It cannot, therefore, alone acquire any further protection through the agency of natural selection. But the female requires some extra protection, to balance the greater danger to which she is exposed, and her greater importance to the existence of the species; and this she always acquires, in one way or another, through the action of natural selection.

In his "Origin of Species," fourth edition, p. 241, Mr. Darwin recognises the necessity for protection as sometimes being a cause of the obscure colours of female birds; but he does not seem to consider it so very important an agent in modifying colour as I am disposed to do. In the same paragraph (p. 240), he alludes to the fact of female birds and butterflies being sometimes very plain, sometimes as gay as the males; but, apparently, considers this mainly due to peculiar laws of inheritance, which sometimes continue acquired colour in the line of one sex only, sometimes in both. Without denying the action of such a law (which Mr.

Darwin informs me he has facts to support), I impute the difference, in the great majority of cases, to the greater or less need of protection in the female sex in these groups of animals.

This need was seen to exist a century ago by the Hon. Daines Barrington, who, in the article already quoted (see p. 220), after alluding to the fact that singing birds are all small, and suggesting (but I think erroneously) that this may have arisen from the difficulty larger birds would have in concealing themselves if they called the attention of their enemies by loud notes, goes on thus:—" I should rather conceive it is for the same reason no hen bird sings, because this talent would be still more dangerous during incubation, which *may possibly also account for the inferiority in point of plumage.*" This is a curious anticipation of the main idea on which this essay is founded. It has been unnoticed for near a century, and my attention was only recently called to it by Mr. Darwin himself.

Conclusion.

To some persons it will perhaps appear, that the causes to which I impute so much of the external aspect of nature are too simple, too insignificant, and too unimportant for such a mighty work. But I would ask them to consider, that the great object of all the peculiarities of animal structure is to preserve the life of the individual, and to maintain the existence of the species. Colour has hitherto been too often looked upon as something adventitious and superficial, something given to an animal not to be useful to itself, but solely to gratify man or even superior beings — to add to the beauty and ideal harmony of nature. .If this were the case, then, it is evident that the colours of organised beings would be an exception to most other natural phenomena. They would not be the product of general laws, or determined by ever-changing external conditions ; and we must give up all enquiry into their origin and causes, since (by the hypothesis) they are dependent on a Will whose motives must ever be unknown to us.   But, strange to say, no sooner do we begin to examine and classify the colours of natural objects, than we find that they are intimately related to a variety of other phenomena, and are, like them, strictly subordinated to general laws.   I have here attempted to elucidate some of these laws in the case of birds, and have shown how the mode of nidification has affected the colouring of the female sex in this group.   I have before shown to how great an extent, and in how many ways, the need of protection has determined the colours of insects, and of some groups of reptiles and mammalia, and I would now call particular attention to the fact that the gay tints of flowers, so long supposed to be a convincing proof that. colour has been bestowed for other purposes than the good of its possessor, have been shown by Mr. Darwin to follow the same great law of utility. Flowers do not often need protection, but very often require the aid of insects to fertilize them, and maintain their reproductive powers in the greatest vigour. Their gay colours attract insects, as do also their sweet odours and honeyed secretions; and that this is the main function of colour in flowers is shown by the striking fact, that those flowers which can be perfectly fertilized by the wind, and do not need the aid of insects, *rarely or never have gaily-coloured flowers.*

This wide extension of the general principle of utility to the colours of such varied groups, both in the animal and vegetable kingdoms, compels us to acknowledge that the " reign of law " has been fairly traced into this stronghold of the advocates of special creation. And to those who oppose the explanation I have given of the facts adduced in this essay, I would again respectfully urge that they must grapple with the whole of the facts, not one or two of them only. It will be admitted that, on the theory of evolution and natural selection, a wide range of facts with regard to colour in nature have been co-ordinated and explained. Until at least an equally wide range of facts can be shown to be in harmony with any other theory, we can hardly be expected to abandon that which has already done such good service, and which has led to the discovery of so many interesting and unexpected harmonies among the most common (but hitherto most neglected and least understood), of the phenomena presented by organised beings.

VIII.

CREATION BY LAW.

AMONG the various criticisms that have appeared on Mr. Darwin's celebrated " Origin of Species," there is, perhaps, none that will appeal to so large a number of well educated and intelligent persons, as that contained in the Duke of Argyll's " Reign of Law." The noble author represents the feelings and expresses the ideas of that large class, who take a keen interest in the progress of Science in general, and especially that of Natural History, but have never themselves studied nature in detail, or acquired that personal knowledge of the structure of closely allied forms,—the wonderful gradations from species to species and from group to group, and the infinite variety of the phenomena of " variation " in organic beings,—which are absolutely necessary for a full appreciation of the facts and reasonings contained in Mr. Darwin's great work.

Nearly half of the Duke's book is devoted to an exposition of his idea of " Creation by Law," and he expresses so clearly what are his difficulties and objections as regards the theory of " Natural Selection," that I think it advisable that they should be fairly answered, and that his own views should be shown to lead to conclusions, as hard to accept as any which he imputes to Mr. Darwin.

The point on which the Duke of Argyll lays most stress, is, that proofs of Mind everywhere meet us in Nature, and are more especially manifest wherever we find " contrivance " or " beauty." He maintains that this indicates the constant supervision and direct interference of the Creator, and cannot possibly be explained by the unassisted action of any combination of laws. Now, Mr. Darwin's work has for its main object, to show, that all the phenomena of living things,—all their wonderful organs and complicated structures, their infinite variety of form, size, and colour, their intricate and involved relations to each other,—may have been produced by the action of a few general laws of the simplest kind, laws which are in most cases mere statements of admitted facts. The chief of these laws or facts are the following:—

1. *The Law of Multiplication in Geometrical Progression.*—All organized beings have enormous powers of multiplication. Even man, who increases slower than all other animals, could under the most favourable circumstances double his numbers every fifteen years, or a hundred-fold in a century. Many animals and plants could increase their numbers from ten to a thousand-fold every year.

2. *The Law of Limited Populations.*—The number of living individuals of each species in any country, or in the whole globe, is practically stationary; whence it follows that the whole of this enormous increase must die off almost as fast as produced, except only those individuals for whom room is made by the death

of parents. As a simple but striking example, take an oak forest. Every oak will drop annually thousands or millions of acorns, but till an old tree falls, not one of these millions can grow up into an oak. They must die at various stages of growth.

3. *The Law of Heredity, or Likeness of Offspring to their Parents.*—This is a universal, but not an absolute law. All creatures resemble their parents in a high degree, and in the majority of cases very accurately; so that even individual peculiarities, of whatever kind, in the parents, are almost always transmitted to some of the offspring.

4. *The Law of Variation.*—This is fully expressed by the lines :—

> " No being on this earthly ball,
> Is like another, all in all."

Offspring resemble their parents very much, but not wholly—each being possesses its individuality. This " variation " itself varies in amount, but it is always present, not only in the whole being, but in every part of every being. Every organ, every character, every feeling is individual ; that is to say, *varies* from the same organ, character, or feeling in every other individual.

5. *The Law of unceasing Change of Physical Conditions upon the Surface of the Earth.*—Geology shows us that this change has always gone on in times past, and we also know that it is now everywhere going on.

6. *The Equilibrium or Harmony of Nature.*—When a species is well adapted to the conditions which

environ it, it flourishes; when imperfectly adapted it decays; when ill-adapted it becomes extinct. If *all* the conditions which determine an organism's well-being are taken into consideration, this statement can hardly be disputed.

This series of facts or laws, are mere statements of what is the condition of nature. They are facts or inferences which are generally known, generally admitted—but in discussing the subject of the " Origin of Species "—as generally forgotten. It is from these universally admitted facts, that the origin of all the varied forms of nature may be deduced by a logical chain of reasoning, which, however, is at every step verified and shown to be in strict accord with facts; and, at the same time, many curious phenomena which can by no other means be understood, are explained and accounted for. It is probable, that these primary facts or laws are but results of the very nature of life, and of the essential properties of organized and unorganized matter. Mr. Herbert Spencer, in his "First Principles" and his "Biology" has, I think, made us able to understand how this may be; but at present we may accept these simple laws without going further back, and the question then is—whether the variety, the harmony, the contrivance, and the beauty we perceive in organic beings, can have been produced by the action of these laws alone, or whether we are required to believe in the incessant interference and direct action of the mind and will of the Creator. It is simply a question of how the Creator has worked. The Duke (and I quote him as having well expressed the views of the more intelligent of Mr. Darwin's opponents) maintains, that He has personally applied general laws to produce effects, which those laws are not in themselves capable of producing; that the universe alone, with all its laws intact, would be a sort of chaos, without variety, without harmony, without design, without beauty; that there is not (and therefore we may presume that there could not be) any self-developing power in the universe. I believe, on the contrary, that the universe is so constituted as to be self-regulating; that as long as it contains Life, the forms under which that life is manifested have an inherent power of adjustment to each other and to surrounding nature; and that this adjustment necessarily leads to the greatest amount of variety and beauty and enjoyment, because it does depend on general laws, and not on a continual supervision and re-arrangement of details. As a matter of feeling and religion, I hold this to be a far higher conception of the Creator and of the Universe that that which may be called the "continual interference" hypothesis; but it is not a question to be decided by our feelings or convictions, it is a question of facts and of reason. Could the change, which Geology shows us has ever taken place in the forms of life, have been produced by general laws, or does it imperatively require the incessant supervision of a creative mind? This is the question for us to consider, and our opponents have the difficult task of proving a negative, if we show that there are both facts and analogies in our favour.

Mr. Darwin's Metaphors liable to Misconception.

Mr. Darwin has laid himself open to much misconception, and has given to his opponents a powerful weapon against himself, by his continual use of metaphor in describing the wonderful co-adaptations of organic beings.

" It is curious," says the Duke of Argyll, "to observe the language which this most advanced disciple of pure naturalism instinctively uses, when he has to describe the complicated structure of this curious order of plants (the Orchids). 'Caution in ascribing intentions to nature,' does not seem to occur to him as possible. Intention is the one thing which he does see, and which, when he does not see, he seeks for diligently until he finds it. He exhausts every form of words and of illustration, by which intention or mental purpose can be described. 'Contrivance'—'curious contrivance,'—'beautiful contrivance,'—these are expressions which occur over and over again. Here is one sentence describing the parts of a particular species : ' the Labellum is developed into a long nectary, *in order* to attract Lepidoptera, and we shall presently give reason for suspecting that the nectar is *purposely* so lodged, that it can be sucked only slowly *in order* to give time for the curious chemical quality of the viscid matter setting hard and dry.'" Many other examples of similar expressions are quoted by the Duke, who maintains that no explanation of these " contrivances " has been or can be given, except on the supposition of a personal contriver, specially arranging the details of each case, although causing them to be produced by the ordinary processes of growth and reproduction.

Now there is a difficulty in this view of the origin of the structure of Orchids which the Duke does not allude to. The majority of flowering plants are fertilized, either without the agency of insects or, when insects are required, without any very important modification of the structure of the flower. It is evident, therefore, that flowers might have been formed as varied, fantastic, and beautiful as the Orchids, and yet have been fertilized without more complexity of structure than is found in Violets, or Clover, or Primroses, or a thousand other flowers. The strange springs and traps and pitfalls found in the flowers of Orchids cannot be necessary *per se*, since exactly the same end is gained in ten thousand other flowers which do not possess them. Is it not then an extraordinary idea, to imagine the Creator of the Universe *contriving* the various complicated parts of these flowers, as a mechanic might contrive an ingenious toy or a difficult puzzle? Is it not a more worthy conception that they are some of the results of those general laws which were so co-ordinated at the first introduction of life upon the earth as to result necessarily in the utmost possible development of varied forms?

But let us take one of the simpler cases adduced and see if our general laws are unable to account for it.

A Case of Orchis-structure explained by Natural Selection.

There is a Madagascar Orchis—the Angræcum ses-quipedale—with an immensely long and deep nectary. How did such an extraordinary organ come to be developed ? Mr. Darwin's explanation is this. The pollen of this flower can only be removed by the base of the proboscis of some very large moths, when trying to get at the nectar at the bottom of the vessel. The moths with the longest probosces would do this most effectually ; they would be rewarded for their long tongues by getting the most nectar ; whilst on the other hand, the flowers with the deepest nectaries would be the best fertilized by the largest moths preferring them. Consequently, the deepest nectaried Orchids and the longest tongued moths would each confer on the other an advantage in the battle of life. This would tend to their respective perpetuation, and to the constant lengthening of nectaries and probosces. Now let it be remembered, that what we have to account for, is only the unusual length of this organ. A nectary is found in many orders of plants and is especially common in the Orchids, but in this one case only is it more than a foot long. How did this arise ? We begin with the fact, proved experimentally by Mr. Darwin, that moths do visit Orchids, do thrust their spiral trunks into the nectaries, and do fertilize them by carrying the pollinia of one flower to the stigma of another. He has further explained the exact mechanism

by which this is effected, and the Duke of Argyll admits the accuracy of his observations. In our British
species, such as Orchis pyramidalis, it is not necessary
that there should be any exact adjustment between
the length of the nectary and that of the proboscis of
the insect; and thus a number of insects of various
sizes are found to carry away the pollinia and aid in
the fertilization. In the Angræcum sesquipedale, however, it is necessary that the proboscis should be forced
into a particular part of the flower, and this would
only be done by a large moth burying its proboscis
to the very base, and straining to drain the nectar
from the bottom of the long tube, in which it occupies
a depth of one or two inches only. Now let us start
from the time when the nectary was only half its
present length or about six inches, and was chiefly
fertilized by a species of moth which appeared at the
time of the plant's flowering, and whose proboscis was
of the same length. Among the millions of flowers
of the Angræcum produced every year, some would
always be shorter than the average, some longer. The
former, owing to the structure of the flower, would
not get fertilized, because the moths could get all the
nectar without forcing their trunks down to the very
base. The latter would be well fertilized, and the
longest would on the average be the best fertilized of
all. By this process alone the average length of the
nectary would annually increase, because, the short-nec-
taried flowers being sterile and the long ones having
abundant offspring, exactly the same effect would be

produced as if a gardener destroyed the short ones and sowed the seed of the long ones only; and this we know by experience would produce a regular increase of length, since it is this very process which has increased the size and changed the form of our cultivated fruits and flowers.

But this would lead in time to such an increased length of the nectary that many of the moths could only just reach the surface of the nectar, and only the few with exceptionally long trunks be able to suck up a considerable portion.

This would cause many moths to neglect these flowers because they could not get a satisfying supply of nectar, and if these were the only moths in the country the flowers would undoubtedly suffer, and the further growth of the nectary be checked by exactly the same process which had led to its increase. But there are an immense variety of moths, of various lengths of proboscis, and as the nectary became longer, other and larger species would become the fertilizers, and would carry on the process till the largest moths became the sole agents. Now, if not before, the moth would also be affected, for those with the longest probosces would get most food, would be the strongest and most vigorous, would visit and fertilize the greatest number of flowers, and would leave the largest number of·descendants. The flowers most completely fertilized by these moths being those which had the longest nectaries, there would in each generation be on the average an increase in the length of the nectaries, and also

T

an average increase in the length of the probosces of
the moths; and this would be a *necessary result* from
the fact that nature ever fluctuates about a mean, or
that in every generation there would be flowers with
longer and shorter nectaries, and moths with longer
and shorter probosces than the average. No doubt
there are a hundred causes that might have checked
this process before it had reached the point of development at which we find it.· If, for instance, the
variation in the quantity of nectar had been at any
stage greater than the variation in the length of the
nectary, then smaller moths could have reached it
and have effected the fertilization. Or if the growth
of the probosces of the moths had from other causes
increased quicker than that of the nectary, or if the
increased length of proboscis had been injurious to
them in any way, or if the species of moth with the
longest proboscis had become much diminished by
some enemy or other unfavourable conditions, then,
in any of these cases, the shorter nectaried flowers,
which would have attracted and could have been fertilized by the smaller kinds of moths, would have had
the advantage. And checks of a similar nature to
these no doubt have acted in other parts of the world,
and have prevented such an extraordinary development of nectary as has been produced by favourable
conditions in Madagascar only, and in one single species
of Orchid. I may here mention that some of the
large Sphinx moths of the tropics have probosces
nearly as long as the nectary of Angræcum sesquipe-

dale. I have carefully measured the proboscis of a specimen of Macrosila cluentius from South America, in the collection of the British Museum, and find it to be nine inches and a quarter long! One from tropical Africa (Macrosila morganii) is seven inches and a half. A species having a proboscis two or three inches longer could reach the nectar in the largest flowers of Angræcum sesquipedale, whose nectaries vary in length from ten to fourteen inches. That such a moth exists in Madagascar may be safely predicted; and naturalists who visit that island should search for it with as much confidence as Astronomers searched for the planet Neptune,—and I venture to predict they will be equally successful!

Now, instead of this beautiful self-acting adjustment, the opposing theory is, that the Creator of the Universe, by a direct act of his Will, so disposed the natural forces influencing the growth of this one species of plant as to cause its nectary to increase to this enormous length; and at the same time, by an equally special act, determined the flow of nourishment in the organization of the moth, so as to cause its proboscis to increase in exactly the same proportion, having previously so constructed the Angræcum that it could only be maintained in existence by the agency of this moth. But what proof is given or suggested that this was the mode by which the adjustment took place? None whatever, except a feeling that there is an adjustment of a delicate kind, and an inability to see how known causes could have produced such an adjustment. I believe I have shown, however, that such an adjustment is not only possible but inevitable, unless at some point or other we deny the action of those simple laws which we have already admitted to be but the expressions of existing facts.

Adaptation brought about by General Laws.

It is difficult to find anything like parallel cases in inorganic nature, but that of a river may perhaps illustrate the subject in some degree. Let us suppose a person totally ignorant of Modern Geology to study carefully a great River System. He finds in its lower part, a deep broad channel filled to the brim, flowing slowly through a flat country and carrying out to the sea a quantity of fine sediment. Higher up it branches into a number of smaller channels, flowing alternately through flat valleys and between high banks; sometimes he finds a deep rocky bed with perpendicular walls, carrying the water through a chain of hills; where the stream is narrow he finds it deep, where wide shallow. Further up still, he comes to a mountainous region, with hundreds of streams and rivulets, each with its tributary rills and gullies, collecting the water from every square mile of surface, and every channel adapted to the water that it has to carry. He finds that the bed of every branch, and stream, and rivulet, has a steeper and steeper slope as it approaches its sources, and is thus enabled to carry off the water from heavy rains, and to bear away

the stones and pebbles and gravel, that would otherwise block up its course.   In every part of this system
he would see exact adaptation of means to an end.
He would say, that this system of channels must have
been designed, it answers its purpose so effectually.
Nothing but a mind could have so exactly adapted
the slopes of the channels, their capacity, and fre-
quency, to the nature of the soil and the quantity of
the rainfall.   Again, he would see special adaptation
to the wants of man, in broad quiet navigable rivers
flowing through fertile plains that support a large
population, while the rocky streams and mountain
torrents, were confined to those sterile regions suitable only for a small population of shepherds and
herdsmen.   He would listen with incredulity to the
Geologist, who assured him, that the adaptation and
adjustment he so admired was an inevitable result of
the action of general laws.   That the rains and rivers,
aided by subterranean forces, had modelled the country,
had formed the hills and valleys, had scooped out the
river beds, and levelled the plains ;—and it would only
be after much patient observation and study, after
having watched the minute changes produced year
by year, and multiplying them by thousands and ten
thousands, after visiting the various regions of the
earth and seeing the changes everywhere going on,
and the unmistakable signs of greater changes in past
times, — that he could be made to understand that
the surface of the earth, however beautiful and harmonious it may appear, is strictly due in every detail

to the action of forces which are demonstrably self-adjusting.

Moreover, when he had sufficiently extended his inquiries, he would find, that every evil effect which he would imagine must be the result of non-adjustment does somewhere or other occur, only it is not always evil. Looking on a fertile valley, he would perhaps say—" If the channel of this river were not well adjusted, if for a few miles it sloped the wrong way, the water could not escape, and all this luxuriant valley, full of human beings, would become a waste of waters." Well, there are hundreds of such cases. Every lake is a valley " wasted by water," and in some cases (as the Dead Sea) it is a positive evil, a blot upon the harmony and adaptation of the surface of the earth. Again, he might say—" If rain did not fall here, but the clouds passed over us to some other regions, this verdant and highly cultivated plain would become a desert." And there are such deserts over a large part of the earth, which abundant rains would convert into pleasant dwelling-places for man. Or he might observe some great navigable river, and reflect how easily rocks, or a steeper channel in places, might render it useless to man ;—and a little inquiry would show him hundreds of rivers in every part of the world, which are thus rendered useless for navigation.

Exactly the same thing occurs in organic nature. We see some one wonderful case of adjustment, some unusual development of an organ, but we pass over the

hundreds of cases in which that adjustment and development do not occur. No doubt when one adjustment is
absent another takes its place, because no organism can
continue to exist that is not adjusted to its environment; and unceasing variation with unlimited powers
of multiplication, in most cases, furnish the means
of self-adjustment. The world is so constituted, that
by the action of general laws there is produced the
greatest possible variety of surface and of climate;
and by the action of laws equally general, the greatest
possible variety of organisms have been produced,
adapted to the varied conditions of every part of the
earth. The objector would probably himself admit,
that the varied surface of the earth—the plains and
valleys, the hills and mountains, the deserts and vol-
canoes, the winds and currents, the seas and lakes
and rivers, and the various climates of the earth—are
all the results of general laws acting and re-acting
during countless ages; and that the Creator does not
appear to guide and control the action of these laws
—here determining the height of a mountain, there
altering the channel of a river—here making the rains
more abundant, there changing the direction of a
current. He would probably admit that the forces of
inorganic nature are self-adjusting, and that the result
necessarily fluctuates about a given mean condition
(which is itself slowly changing), while within certain
limits the greatest possible amount of variety is produced. If then a "contriving mind" is not necessary at every step of the process of change eternally

going on in the inorganic world, why are we required to believe in the continual action of such
a mind in the region of organic nature? True, the
laws at work are more complex, the adjustments more
delicate, the appearance of special adaptation more remarkable; but why should we measure the creative
mind by our own? Why should we suppose the machine too complicated, to have been designed by the
Creator so complete that it would necessarily work out
harmonious results? The theory of " continual interference" is a limitation of the Creator's power. It
assumes that he could not work by pure law in the
organic, as he has done in the inorganic world; it
assumes that he could not foresee the consequences
of the laws of matter and mind combined—that results would continually arise which are contrary to
what is best, and that he has to change what would
otherwise be the course of nature, in order to produce
that beauty, and variety, and harmony, which even we,
with our limited intellects, can conceive to be the
result of self-adjustment in a universe governed by
unvarying law. If we could not conceive the world
of nature to be self-adjusting and capable of endless
development, it would even then be an unworthy idea
of a Creator, to impute the incapacity of our minds
to him; but when many human minds can conceive,
and can even trace out in detail some of the adaptations in nature as the necessary results of unvarying
law, it seems strange that, in the interests of religion,
any one should seek to prove that the System of Na-

ture, instead of being above, is far below our highest conceptions of it. I, for one, cannot believe that the world would come to chaos if left to Law alone. I cannot believe that there is in it no inherent power of developing beauty or variety, and that the direct action of the Deity is required to produce each spot or streak on every insect, each detail of structure in every one of the millions of organisms that live or have lived upon the earth. For it is impossible to draw a line. If any modifications of structure could be the result of law, why not all? If some self-adaptations could arise, why not others? If any varieties of colour, why not all the varieties we see? No attempt is made to explain this, except by reference to the fact that "purpose" and "contrivance" are everywhere visible, and by the illogical deduction that they could only have arisen from the direct action of some mind, because the direct action of our minds produces similar "contrivances"; but it is forgotten that adaptation, however produced, must have the appearance of design. The channel of a river looks as if made *for* the river, although it is made *by* it; the fine layers and beds in a deposit of sand, often look as if they had been sorted, and sifted, and levelled, designedly; the sides and angles of a crystal exactly resemble similar forms designed by man; but we do not therefore conclude that these effects have, in each individual case, required the directing action of a creative mind, or see any difficulty in their being produced by natural Law.

Beauty in Nature.

Let us, however, leave this general argument for a while, and turn to another special case, which has been appealed to as conclusive against Mr. Darwin's views. " Beauty " is, to some persons, as great a stumbling-block as " contrivance." They cannot conceive a system of the Universe, so perfect, as necessarily to develop every form of Beauty, but suppose that when anything specially beautiful occurs, it is a step beyond what that system could have produced, something which the Creator has added for his own delectation.

Speaking of the Humming Birds, the Duke of Argyll says: " In the first place, it is to be observed of the whole group, that there is no connection which can be traced or conceived, between the splendour of the humming birds and any function essential to their life. If there were any such connection, that splendour could not be confined, as it almost exclusively is, to only one sex. The female birds are, of course, not placed at any disadvantage in the struggle for existence by their more sombre colouring." And after describing the various ornaments of these birds, he says: " Mere ornament and variety of form, and these for their own sake, is the only principle or rule with reference to which Creative Power seems to have worked in these wonderful and beautiful birds. . . A crest of topaz is no better in the struggle for existence than a crest of sapphire. A frill ending in spangles of the emerald is no better in the battle of life than a frill ending in spangles of the ruby. A tail is not affected for the purposes of flight, whether its marginal or its central feathers are decorated with white. . . Mere beauty and mere variety, for their own sake, are objects which we ourselves seek when we can make the Forces of Nature subordinate to the attainment of them. There seems to be no conceivable reason why we should doubt or question, that these are ends and aims also in the forms given to living organisms " (" Reign of Law," p. 248).

Here the statement that " no connection can be conceived between the splendour of the humming birds and any function essential to their life," is met by the fact, that Mr. Darwin has not only conceived but has shown, both by observation and reasoning, how beauty of colour and form may have a direct influence on the most important of all the functions of life, that of reproduction. In the variations to which birds are subject, any more brilliant colour than usual would be attractive to the females, and would lead to the individuals so adorned leaving more than the average number of offspring. Experiment and observation have shown, that this kind of sexual selection does actually take place; and the laws of inheritance would necessarily lead to the further development of any individual peculiarity that was attractive, and thus the splendour of the humming birds is directly connected with their very existence. It is true that " a crest of topaz may be no better than a

crest of sapphire," but either of these may be much
better than no crest at all; and the different conditions
under which the parent form must have existed in
different parts of its range, will have determined different variations of tint, either of which were advantageous.   The reason why female birds are not
adorned with equally brilliant plumes is sufficiently
clear; they would be injurious, by rendering their pos-
sessors too conspicuous during incubation.   Survival
of the fittest, has therefore favoured the development
of those dark green tints on the upper surface of so
many female humming birds, which are most conducive
to their protection while the important functions of
hatching and rearing the young are being carried on.
Keeping in mind the laws of multiplication, variation,
and survival of the fittest, which are for ever in action,
these varied developments of beauty and harmonious
adjustments to conditions, are not only conceivable
but demonstrable results.

The objection I am now combating is solely founded
on the supposed analogy of the Creator's mind to
ours, as regards the love of Beauty for its own sake;
but if this analogy is to be trusted, then there ought
to be no natural objects which are disagreeable or
ungraceful in our eyes.   And yet it is undoubtedly
the fact that there are many such.   Just as surely
as the Horse and Deer are beautiful and graceful,
the Elephant, Rhinoceros, Hippopotamus, and Camel
are the reverse.   The majority of Monkeys and Apes
are not beautiful; the majority of Birds have no beauty

of colour; a vast number of Insects and Reptiles are positively ugly. Now, if the Creator's mind is like ours, whence this ugliness? It is useless to say "that is a mystery we cannot explain," because we have attempted to explain one-half of creation by a method that will not apply to the other half. We know that a man with the highest taste and with unlimited wealth, practically does abolish all ungraceful and disagreeable forms and colours from his own domains. If the beauty of creation is to be explained by the Creator's love of beauty, we are bound to ask why he has not banished deformity from the earth, as the wealthy and enlightened man does from his estate and from his dwelling; and if we can get no satisfactory answer, we shall do well to reject the explanation offered. Again, in the case of flowers, which are always especially referred to, as the surest evidence of beauty being an end of itself in creation, the whole of the facts are never fairly met. At least half the plants in the world have not bright-coloured or beautiful flowers; and Mr. Darwin has lately arrived at the wonderful generalization, that flowers have become beautiful solely to attract insects to assist in their fertilization. He adds, "I have come to this conclusion from finding it an invariable rule, that when a flower is fertilized by the wind it never has a gaily-coloured corolla." Here is a most wonderful case of beauty being *useful*, when it might be least expected. But much more is proved; for when beauty is of no use to the plant it is not given. It cannot be imag-

ined to do any harm.   It is simply not necessary, and
is therefore withheld !   We ought surely to have been
told how this fact is consistent with beauty being " an
end in itself," and 'with the statement of its being
given to natural objects " for its own sake."

How new Forms are produced by Variation and Selection.

Let us now consider another of the popular objections which the Duke of Argyll thus sets forth :—

" Mr. Darwin does not pretend to have discovered
any law or rule, according to which new Forms have
been born from old Forms. He does not hold that
outward conditions, however changed, are sufficient to
account for them. . . His theory seems to be far
better than a mere theory—to be an established scientific truth—in so far as it accounts, in part at least,
for the success and establishment and spread of new
Forms *when they have arisen.* But it does not even
suggest the law under which, or by or according to
which, such new Forms are introduced. Natural Selection can do nothing, except with the materials
presented to its hands. It cannot select except among
the things open to selection. . . Strictly speaking,
therefore, Mr. Darwin's theory is not a theory on
the Origin of Species at all, but only a theory on the
causes which lead to the relative success or failure
of such new forms as may be born into the world."
(" Reign of Law," p. 230.)

In this, and many other passages in his work, the

Duke of Argyll sets forth his idea of Creation as a " Creation by birth," but maintains that each birth of a new form from parents differing from itself, has been produced by a special interference of the Creator, in order to direct the process of development into certain channels ; that each new species is in fact a " special creation," although brought into existence through the ordinary laws of reproduction. He maintains therefore, that the laws of multiplication and variation cannot furnish the right kinds of materials at the right times for natural selection to work on. I believe, on the contrary, that it can be logically *proved* from the six axiomatic laws before laid down, that such materials would be furnished ; but I prefer to show there are abundance of *facts* which demonstrate that they are furnished.

The experience of all cultivators of plants and breeders of animals shows, that when a sufficient number of individuals are examined, variations of any required kind can always be met with. On this depends the possibility of obtaining breeds, races, and fixed varieties of animals and plants; and it is found, that any one form of variation may be accumulated by selection, without materially affecting the other characters of the species; each *seems* to vary in the one required direction only. For example, in turnips, radishes, potatoes, and carrots, the root or tuber varies in size, colour, form, and flavour, while the foliage and flowers seem to remain almost stationary ; in the cabbage and lettuce, on the contrary, the foliage can be modified into various forms and modes of growth, the root, flower, and fruit remaining little altered; in the cauliflower and brocoli the flower heads vary; in the garden pea the pod only changes. We get innumerable forms of fruit in the apple and pear, while the leaves and flowers remain undistinguishable; the same occurs in the gooseberry and garden currant. Directly however, (in the very same genus) we want the flower to vary in the Ribes sanguineum, it does so, although mere cultivation for hundreds of years has not produced marked differences in the flowers of Ribes grossularia. When fashion demands any particular change in the form or size, or colour of a flower, sufficient variation always occurs in the right direction, as is shown by our roses, auriculas, and geraniums; when, as recently, ornamental leaves come into fashion sufficient variation is found to meet the demand, and we have zoned pelargoniums, and variegated ivy, and it is discovered that a host of our commonest shrubs and herbaceous plants have taken to vary in this direction just when we want them to do so! This rapid variation is not confined to old and well-known plants subjected for a long series of generations to cultivation, but the Sikim Rhododendrons, the Fuchsias, and Calceolarias from the Andes, and the Pelargoniums from the Cape are equally accommodating, and vary just when and where and how we require them.

Turning to animals we find equally striking examples. If we want any special quality in any animal we have only to breed it in sufficient quantities and watch carefully, and the required variety is *always* found, and can be increased to almost any desired extent. In Sheep, we get flesh, fat, and wool; in Cows, milk; in Horses, colour, strength, size, and speed; in Poultry, we have got almost any variety of colour, curious modifications of plumage, and the capacity of perpetual egg-laying. In Pigeons we have a still more remarkable proof of the universality of variation, for it has been at one time or another the fancy of breeders to change the form of every part of these birds, and they have never found the required variations absent. The form, size, and shape of bill and feet, have been changed to such a degree as is found only in distinct genera of wild birds; the number of tail feathers has been increased, a character which is generally one of the most permanent nature, and is of high importance in the classification of birds; and the size, the colour, and the habits, have been also changed to a marvellous extent. In Dogs, the degree of modification and the facility with which it is effected, is almost equally apparent. Look at the constant amount of variation in opposite directions that must have been going on, to develop the poodle and the greyhound from the same original stock! Instincts, habits, intelligence, size, speed, form, and colour, have always varied, so as to produce the very races which the wants or fancies or passions of men may have led them to desire. Whether they wanted a bull-dog to torture another animal, a greyhound to catch a hare, or a bloodhound to hunt down their oppressed fellow-creatures, the required variations have always appeared.

Now this great mass of facts, of which a mere sketch has been here given, are fully accounted for by the " Law of Variation " as laid down at the commencement of this paper. Universal variability— small in amount but in every direction, ever fluctuating about a mean condition until made to advance in a given direction by " selection," natural or artificial, —is the simple basis for the indefinite modification of the forms of life;—partial, unbalanced, and consequently unstable modifications being produced by man, while those developed under the unrestrained action of natural laws, are at every step self-adjusted to external conditions by the dying out of all unadjusted forms, and are therefore stable and comparatively permanent. To be consistent in their views, our opponents must maintain that every one of the variations that have rendered possible the changes produced by man, have been determined at the right time and place by the will of the Creator. Every race produced by the florist or the breeder, the dog or the pigeon fancier, the ratcatcher, the sporting man, or the slave-hunter, must have been provided for by varieties occurring when wanted; and as these variations were never withheld, it would prove, that the sanction of an all-wise and all-powerful Being, has been given to that which the highest human minds consider to be trivial, mean, or debasing.

This appears to be a complete answer to the theory, that variation sufficient in amount to be accumulated in a given direction must be the direct act of the Creative Mind, but it is also sufficiently condemned by being so entirely unnecessary. The facility with which man obtains new races, depends chiefly upon the number of individuals he can procure to select from. When hundreds of florists or breeders are all aiming at the same object, the work of change goes on rapidly. But a common species in nature contains a thousand- or a million-fold more individuals than any domestic race; and survival of the fittest must unerringly preserve all that vary in the right direction, not only in obvious characters but in minute details, not only in external but in internal organs; so that if the materials are sufficient for the needs of man, there can be no want of them to fulfil the grand purpose of keeping up a supply of modified organisms, exactly adapted to the changed conditions that are always occurring in the inorganic world.

The Objection that there are Limits to Variation.

Having now, I believe, fairly answered the chief objections of the Duke of Argyll, I proceed to notice one or two of those adduced in an able and argumentative essay on the " Origin of Species " in the *North British Review* for July, 1867. The writer first attempts to prove that there are strict limits to variation. When we begin to select variations in any one direction, the process is comparatively rapid, but after a considerable

amount of change has been effected it becomes slower
and slower, till at length its limits are reached and no
care in breeding and selection can produce any further
advance. The race-horse is chosen as an example.
It is admitted that, with any ordinary lot of horses
to begin with, careful selection would in a few years
make a great improvement, and in a comparatively
short time the standard of our best racers might be
reached. But that standard has not for many years
been materially raised, although unlimited wealth and
energy are expended in the attempt. This is held to
prove that there are definite limits to variation in any
special direction, and that we have no reason to suppose that mere time, and the selective process being
carried on by natural law, could make any material
difference. But the writer does not perceive that this
argument fails to meet the real question, which is, not
whether indefinite and unlimited change in any or all
directions is possible, but whether such differences as
do occur in nature could have been produced by the
accumulation of variations by selection. In the matter
of speed, a limit of a definite kind as regards land
animals does exist in nature. All the swiftest animals
—deer, antelopes, hares, foxes, lions, leopards, horses,
zebras, and many others, have reached very nearly the
same degree of speed. Although the swiftest of each
must have been for ages preserved, and the slowest
must have perished, we have no reason to believe
there is any advance of speed. The possible limit
under existing conditions, and perhaps under possible

terrestrial conditions, has been long ago reached. In cases, however, where this limit had not been so nearly reached as in the horse, we have been enabled to make a more marked advance and to produce a greater difference of form. The wild dog is an animal that hunts much in company, and trusts more to endurance than to speed. Man has produced the greyhound, which differs much more from the wolf or the dingo than the racer does from the wild Arabian. Domestic dogs, again, have varied more in size and in form than the whole family of Canidæ in a state of nature. No wild dog, fox, or wolf, is either so small as some of the smallest terriers and spaniels, or so large as the largest varieties of hound or Newfoundland dog. And, certainly, no two wild animals of the family differ so widely in form and proportions as the Chinese pug and the Italian greyhound, or the bulldog and the common greyhound. The known range of variation is, therefore, more than enough for the derivation of all the forms of Dogs, Wolves, and Foxes from a common ancestor.

Again, it is objected that the Pouter or the Fantail pigeon cannot be further developed in the same direction. Variation seems to have reached its limits in these birds. But so it has in nature. The Fantail has not only more tail feathers than any of the three hundred and forty existing species of pigeons, but more than any of the eight thousand known species of birds. There is, of course, some limit to the number of feathers of which a tail useful for flight

can consist, and in the Fan-tail we have probably
reached that limit. Many birds have the œsophagus
or the skin of the neck more or less dilatable, but in
no known bird is it so dilatable as in the Pouter
pigeon. Here again the possible limit, compatible
with a healthy existence, has probably been reached.
In like manner the differences in the size and form
of the beak in the various breeds of the domestic
Pigeon, is greater than that between the extreme
forms of beak in the various genera and sub-families
of the whole Pigeon tribe. From these facts, and
many others of the same nature, we may fairly infer,
that if rigid selection were applied to any organ, we
could in a comparatively short time produce a much
greater amount of change than that which occurs between species and species in a state of nature, since
the differences which we do produce are often comparable with those which exist between distinct genera
or distinct families. The facts adduced by the writer
of the article referred to, of the definite limits to variability in certain directions in domesticated animals,
are, therefore, no objection whatever to the view, that
all the modifications which exist in nature have been
produced by the accumulation, by natural selection, of
small and useful variations, since those very modifications have equally definite and very similar limits.

Objection to the Argument from Classification.

To another of this writer's objections—that by Professor Thomson's calculations the sun can only have

existed in a solid state 500,000,000 of years, and that therefore *time* would not suffice for the slow process of development of all living organisms — it is hardly necessary to reply, as it cannot be seriously contended, even if this calculation has claims to approximate accuracy, that the process of change and development may not have been sufficiently rapid to have occurred within that period. His objection to the Classification argument is, however, more plausible. The uncertainty of opinion among Naturalists as to which are species and which varieties, is one of Mr. Darwin's very strong arguments that these two names cannot belong to things quite distinct in nature and origin. The Reviewer says that this argument is of no weight, because the works of man present exactly the same phenomena; and he instances patent inventions, and the excessive difficulty of determining whether they are new or old. I accept the analogy though it is a very imperfect one, and maintain that such as it is, it is all in favour of Mr. Darwin's views. For are not all inventions of the same kind directly affiliated to a common ancestor? Are not improved Steam Engines or Clocks the lineal descendants of some existing Steam Engine or Clock? Is there ever a new Creation in Art or Science any more than in Nature? Did ever patentee absolutely originate any complete and entire invention, no portion of which was derived from anything that had been made or described before? It is therefore clear that the difficulty of distinguishing the various classes of inventions which claim to be new, is of the same nature as the difficulty of distinguishing varieties and species, because neither are absolute new creations, but both are alike descendants of pre-existing forms, from which and from each other they differ by varying and often imperceptible degrees. It appears, then, that however plausible this writer's objections may seem, whenever he descends from generalities to any specific statement, his supposed difficulties turn out to be in reality strongly confirmatory of Mr. Darwin's view.

The " Times," on Natural Selection.

The extraordinary misconception of the whole subject by popular writers and reviewers, is well shown by an article which appeared in the *Times* newspaper on " The Reign of Law." Alluding to the supposed economy of nature, in the adaptation of each species to its own place and its special use, the reviewer remarks : " To this universal law of the greatest economy, the law of natural selection stands in direct antagonism as the law of ' greatest possible waste ' of time and of creative power. To conceive a duck with webbed feet and a spoon-shaped bill, living by suction, to pass naturally into a gull with webbed feet and a knife-like bill, living on flesh, in the longest possible time and in the most laborious possible way, we may conceive it to pass from the one to the other state by natural selection. The battle of life the ducks will have to fight will increase in peril continually as they cease (with the change of their bill) to be ducks, and attain a *maximum* of danger in the condition in which they begin to be gulls; and ages must elapse and whole generations must perish, and countless generations of the one species be created and sacrificed, to arrive at one single pair of the other."

In this passage the theory of natural selection is so absurdly misrepresented that it would be amusing, did we not consider the misleading effect likely to be produced by this kind of teaching in so popular a journal. It is assumed that the duck and the gull are essential parts of nature, each well fitted for its place, and that if one had been produced from the other by a gradual metamorphosis, the intermediate forms would have been useless, unmeaning, and unfitted for any place, in the system of the universe. Now, this idea can only exist in a mind ignorant of the very foundation and essence of the theory of natural selection, which is, the preservation of *useful* variations only, or, as has been well expressed, in other words, the " survival of the fittest." Every intermediate form which could possibly have arisen during the transition from the duck to the gull, so far from having an unusually severe battle to fight for existence, or incurring any " *maximum* of danger," would necessarily have been as accurately adjusted to the rest of nature, and as well fitted to maintain and to enjoy its existence, as the duck or the gull actually are. If it were not so, it never could have been produced under the law of natural selection.

Intermediate or generalized Forms of extinct Animals,
an indication of Transmutation or Development.

The misconception of this writer illustrates another
point very frequently overlooked. It is an essential
part of Mr. Darwin's theory, that one existing animal
has not been derived from any other existing animal,
but that both are the descendants of a common ancestor, which was at once different from either, but,
in essential characters, intermediate between them both.
The illustration of the duck and the gull is therefore
misleading; one of these birds has not been derived
from the other, but both from a common ancestor.
This is not a mere supposition invented to support the
theory of natural selection, but is founded on a variety
of indisputable facts. As we go back into past time,
and meet with the fossil remains of more and more
ancient races of extinct animals, we find that many
of them actually are intermediate between distinct
groups of existing animals. Professor Owen continually dwells on this fact : he says in his " Palæon-
tology," p. 284 : " A more generalized vertebrate
structure is illustrated, in the extinct reptiles, by
the affinities to ganoid fishes, shown by Ganocephala,
Labyrinthodontia, and Icthyopterygia; by the affinities
of the Pterosauria to Birds, and by the approximation
of the Dinosauria to Mammals. (These have been recently shown by Professor Huxley to have more affinity
to Birds.) It is manifested by the combination of
modern crocodilian, chelonian, and lacertian characters

in the Cryptodontia and the Dicnyodontia, and by the combined lacertian and crocodilian characters in the Thecodontia and Sauropterygia." In the same work he tells us that, "the Anoplotherium, in several important characters resembled the embryo Ruminant, but retained throughout life those marks of adhesion to a generalized mammalian type;"—and assures us that he has "never omitted a proper opportunity for impressing the results of observations showing the more generalized structures of extinct as compared with the more specialized forms of recent animals." Modern palæontologists have discovered hundreds of examples of these more generalized or ancestral types. In the time of Cuvier, the Ruminants and the Pachyderms were looked upon as two of the most distinct orders of animals; but it is now demonstrated that there once existed a variety of genera and species, connecting by almost imperceptible grades such widely different animals as the pig and the camel. Among living quadrupeds we can scarcely find a more isolated group than the genus Equus, comprising the horses, asses, and Zebras; but through many species of Paloplotherium, Hippotherium, and Hipparion, and numbers of extinct forms of Equus found in Europe, India, and America, an almost complete transition is established with the Eocene Anoplotherium and Paleotherium, which are also generalized or ancestral types of the Tapir and Rhinoceros. The recent researches of M. Gaudry in Greece have furnished much new evidence of the same character. In the Miocene beds of Pikermi

he has discovered the group of the Simocyonidæ intermediate between bears and wolves; the genus Hyænictis
which connects the hyænas with the civets; the Ancylo-
therium, which is allied both to the extinct mastodon
and to the living pangolin or scaly ant-eater; and
the Helladotherium, which connects the now isolated
giraffe with the deer and antelopes.

Between reptiles and fishes an intermediate type has
been found in the Archegosaurus of the Coal formation; while the Labyrinthodon of the Trias combined
characters of the Batrachia with those of crocodiles,
lizards, and ganoid fishes.   Even birds, the most apparently isolated of all living forms, and the most rarely
preserved in a fossil state, have been shown to possess
undoubted affinities with reptiles; and in the Oolitic
Archæopteryx, with its lengthened tail, feathered on
each side, we have one of the connecting links from
the side of birds; while Professor Huxley has recently
shown that the entire order of Dinosaurians have remarkable affinities to birds, and that one of them, the
Compsognathus, makes a nearer approach to bird organisation than does Archæopteryx to that of reptiles.

Analogous facts to these occur in other classes of
animals, as an example of which we have the authority
of a distinguished paleontologist, M. Barande, quoted
by Mr. Darwin, for the statement, that although the
Palæozoic Invertebrata can certainly be classed under
existing groups, yet at this ancient period the groups
were not so distinctly separated from each other as
they are now ; while Mr. Scudder tells us, that some of

the fossil insects discovered in the Coal formation of
America offer characters intermediate between those of
existing orders. Agassiz, again, insists strongly that the
more ancient animals resemble the embryonic forms of
existing species; but as the embryos of distinct groups
are known to resemble each other more than the adult
animals (and in fact to be undistinguishable at a very
early age), this is the same as saying that the ancient
animals are exactly what, on Darwin's theory, the
ancestors of existing animals ought to be; and this,
it must be remembered, is the evidence of one of the
strongest opponents of the theory of natural selection.

Conclusion.

I have thus endeavoured to meet fairly, and to answer plainly, a few of the most common objections to
the theory of natural selection, and I have done so in
every case by referring to admitted facts and to logical
deductions from those facts.

As an indication and general summary of the line
of argument I have adopted, I here give a brief demonstration in a tabular form of the Origin of Species
by means of Natural Selection, referring for the *facts*
to Mr. Darwin's works, and to the pages in this volume,
where they are more or less fully treated.

A Demonstration of the Origin of Species by Natural Selection.

PROVED FACTS.	*NECESSARY CONSEQUENCES* (afterwards taken as Proved Facts).
RAPID INCREASE OF ORGANISMS, pp 29, 265; ("Origin of Species," p. 75, 5th Ed.) TOTAL NUMBER OF INDIVIDUALS STATIONARY, pp. 30, 266.	STRUGGLE FOR EXISTENCE, the deaths equalling the births on the average, p. 30; ("Origin of Species," chap. III.)
STRUGGLE FOR EXISTENCE. HEREDITY WITH VARIATION, or general likeness with individual differences of parents and offspring, pp. 266, 287-291, 308; ("Origin of Species," chap. I., II., V.)	SURVIVAL OF THE FITTEST, or Natural Selection; meaning simply, that on the whole those die who are least fitted to maintain their existence; ("Origin of Species," chap. IV.)
SURVIVAL OF THE FITTEST. CHANGE OF EXTERNAL CONDITIONS, universal and unceasing. — See "Lyell's Principles of Geology."	CHANGES OF ORGANIC FORMS, to keep them in harmony with the Changed Conditions; and as the changes of conditions are permanent changes, in the sense of not reverting back to identical previous conditions, the changes of organic forms must be in the same sense permanent, and thus originate SPECIES.

IX.

THE DEVELOPMENT OF HUMAN RACES UNDER THE LAW OF NATURAL SELECTION.

AMONG the most advanced students of man, there exists a wide difference of opinion on some of the most vital questions respecting his nature and origin. Anthropologists are now, indeed, pretty well agreed that man is not a recent introduction into the earth. All who have studied the question, now admit that his antiquity is very great; and that, though we have to some extent ascertained the minimum of time during which he *must* have existed, we have made no approximation towards determining that far greater period during which he *may* have, and probably *has* existed. We can with tolerable certainty affirm that man must have inhabited the earth a thousand centuries ago, but we cannot assert that he positively did not exist, or that there is any good evidence against his having existed, for a period of ten thousand centuries. We know positively, that he was contemporaneous with many now extinct animals, and has survived changes of the earth's surface fifty or a hundred times greater than any that have occurred during the historical period; but we cannot place any definite limit to the number

of species he may have outlived, or to the amount of
terrestrial change he may have witnessed.

Wide differences of opinion as to Man's Origin.

But while on this question of man's antiquity there
is a very general agreement,—and all are waiting
eagerly for fresh evidence to clear up those points
which all admit to be full of doubt,—on other, and
not less obscure and difficult questions, a considerable
amount of dogmatism is exhibited ; doctrines are put
forward as established truths, no doubt or hesitation
is admitted, and it seems to be supposed that no
further evidence is required, or that any new facts
can modify our convictions.   This is especially the case
when we inquire,—Are the various forms under which
man now exists primitive, or derived from pre-existing forms ; in other words, is man of one or many
species ?   To this question we immediately obtain distinct answers diametrically opposed to each other : the
one party positively maintaining, that man is a *species*
and is essentially *one*—that all differences are but local
and temporary variations, produced by the different
physical and moral conditions by which he is surrounded ; the other party maintaining with equal confidence, that man is a genus of *many species*, each of
which is practically unchangeable, and has ever been
as distinct, or even more distinct, than we now behold them.   This difference of opinion is somewhat
remarkable, when we consider that both parties are
well acquainted with the subject ; both use the same

vast accumulation of facts; both reject those early traditions of mankind which profess to give an account of his origin; and both declare that they are seeking fearlessly after truth alone; yet each will persist in looking only at the portion of truth on his own side of the question, and at the error which is mingled with his opponent's doctrine. It is my wish to show how the two opposing views can be combined, so as to eliminate the error and retain the truth in each, and it is by means of Mr. Darwin's celebrated theory of "Natural Selection" that I hope to do this, and thus to harmonise the conflicting theories of modern anthropologists.

Let us first see what each party has to say for itself. In favour of the unity of mankind it is argued, that there are no races without transitions to others; that every race exhibits within itself variations of colour, of hair, of feature, and of form, to such a degree as to bridge over, to a large extent, the gap that separates it from other races. It is asserted that no race is homogeneous; that there is a tendency to vary; that climate, food, and habits produce, and render permanent, physical peculiarities, which, though slight in the limited periods allowed to our observation, would, in the long ages during which the human race has existed, have sufficed to produce all the differences that now appear. It is further asserted that the advocates of the opposite theory do not agree among themselves; that some would make three, some five, some fifty or a hundred and fifty species of man; some would have

x

had each species created in pairs, while others require nations to have at once sprung into existence, and that there is no stability or consistency in any doctrine but that of one primitive stock.

The advocates of the original diversity of man, on the other hand, have much to say for themselves. They argue that proofs of change in man have never been brought forward except to the most trifling amount, while evidence of his permanence meets us everywhere. The Portuguese and Spaniards, settled for two or three centuries in South America, retain their chief physical, mental, and moral characteristics; the Dutch boers at the Cape, and the descendants of the early Dutch settlers in the Moluccas, have not lost the features or the colour of the Germanic races; the Jews, scattered over the world in the most diverse climates, retain the same characteristic lineaments everywhere; the Egyptian sculptures and paintings show us that, for at least 4000 or 5000 years, the strongly contrasted features of the Negro and the Semitic races have remained altogether unchanged; while more recent discoveries prove, that the mound-builders of the Mississippi valley, and the dwellers on Brazilian mountains, had, even in the very infancy of the human race, some traces of the same peculiar and characteristic type of cranial formation that now distinguishes them.

If we endeavour to decide impartially on the merits of this difficult controversy, judging solely by the evidence that each party has brought forward, it certainly seems that the best of the argument is on the side of those who maintain the primitive diversity of man. Their opponents have not been able to refute the permanence of existing races as far back as we can trace them, and have failed to show, in a single case, that at any former epoch the well marked varieties of mankind approximated more closely than they do at the present day. At the same time this is but negative evidence. A condition of immobility for four or five thousand years, does not preclude an advance at an earlier epoch, and—if we can show that there are causes in nature which would check any further physical change when certain conditions were fulfilled— does not even render such an advance improbable, if there are any general arguments to be adduced in its favour. Such a cause, I believe, does exist; and I shall now endeavour to point out its nature and its mode of operation.

Outline of the Theory of Natural Selection.

In order to make my argument intelligible, it is necessary for me to explain very briefly the theory of " Natural Selection " promulgated by Mr. Darwin, and the power which it possesses of modifying the forms of animals and plants. The grand feature in the multiplication of organic life is, that close general resemblance is combined with more or less individual variation. The child resembles its parents or ancestors more or less closely in all its peculiarities, deformities, or beauties ; it resembles them in general more than it

does any other individuals; yet children of the same
parents are not all alike, and it often happens that
they differ very considerably from their parents and
from each other. This is equally true, of man, of all
animals, and of all plants. Moreover, it is found that
individuals do not differ from their parents in certain
particulars only, while in all others they are exact
duplicates of them. They differ from them and from
each other, in every particular: in form, in size, in
colour; in the structure of internal as well as of external
organs; in those subtle peculiarities which produce
differences of constitution, as well as in those still more
subtle ones which lead to modifications of mind and
character. In other words, in every possible way, in
every organ and in every function, individuals of the
same stock vary.

Now, health, strength, and long life, are the results
of a harmony between the individual and the universe
that surrounds it. Let us suppose that at any given
moment this harmony is perfect. A certain animal is
exactly fitted to secure its prey, to escape from its
enemies, to resist the inclemencies of the seasons, and
to rear a numerous and healthy offspring. But a
change now takes place. A series of cold winters, for
instance, come on, making food scarce, and bringing
an immigration of some other animals to compete with
the former inhabitants of the district. The new immigrant is swift of foot, and surpasses its rivals in the
pursuit of game; the winter nights are colder, and
require a thicker fur as a protection, and more

nourishing food to keep up the heat of the system. Our supposed perfect animal is no longer in harmony with its universe; it is in danger of dying of cold or of starvation. But the animal varies in its offspring. Some of these are swifter than others — they still manage to catch food enough; some are hardier and more thickly furred—they manage in the cold nights to keep warm enough; the slow, the weak, and the thinly clad soon die off. Again and again, in each succeeding generation, the same thing takes place. By this natural process, which is so inevitable that it cannot be conceived not to act, those best adapted to live, live; those least adapted, die. It is sometimes said that we have no direct evidence of the action of this selecting power in nature. But it seems to me we have better evidence than even direct observation would be, because it is more universal, viz., the evidence of necessity. It must be so; for, as all wild animals increase in a geometrical ratio, while their actual numbers remain on the average stationary, it follows, that as many die annually as are born. If, therefore, we deny natural selection, it can only be by asserting that, in such a case as I have supposed, the strong, the healthy, the swift, the well clad, the well organised animals in every respect, have no advantage over,—do not on the average live longer than, the weak, the unhealthy, the slow, the ill-clad, and the imperfectly organised individuals; and this no sane man has yet been found hardy enough to assert. But this is not all; for the offspring on the average resemble their parents, and the selected portion of each succeeding generation will therefore be stronger, swifter, and more thickly furred than the last; and if this process goes on for thousands of generations, our animal will have again become thoroughly in harmony with the new conditions in which it is placed. But it will now be a different creature. It will be not only swifter and stronger, and more furry, it will also probably have changed in colour, in form, perhaps have acquired a longer tail, or differently shaped ears; for it is an ascertained fact, that when one part of an animal is modified, some other parts almost always change, as it were in sympathy with it. Mr. Darwin calls this "correlation of growth," and gives as instances, that hairless dogs have imperfect teeth; white cats, when blue-eyed, are deaf; small feet accompany short beaks in pigeons; and other equally interesting cases.

Grant, therefore, the premises: 1st. That peculiarities of every kind are more or less hereditary. 2nd. That the offspring of every animal vary more or less in all parts of their organization. 3rd. That the universe in which these animals live, is not absolutely invariable;—none of which propositions can be denied; and then consider, that the animals in any country (those at least which are not dying out) must at each successive period be brought into harmony with the surrounding conditions; and we have all the elements for a change of form and structure in the animals, keeping exact pace with changes of whatever nature in the surrounding universe. Such changes must be slow, for the changes in the universe are very slow ; but just as these slow changes become important, when we look at results after long periods of action, as we do when we perceive the alterations of the earth's surface during geological epochs; so the parallel changes in animal form become more and more striking, in proportion as the time they have been going on is great ; as we see when we compare our living animals with those which we disentomb from each successively older geological formation.

This is, briefly, the theory of " natural selection," which explains the changes in the organic world as being parallel with, and in part dependent on, those in the inorganic.   What we now have to inquire is,— Can this theory be applied in any way to the question of the origin of the races of man ? or is there anything in human nature that takes him out of the category of those organic existences, over whose successive mutations it has had such powerful sway ?

Different effects of Natural Selection on Animals and on Man.

In order to answer these questions, we must consider why it is that " natural selection " acts so powerfully upon animals ; and we shall, I believe, find, that its effect depends mainly upon their self-dependence and individual isolation.   A slight injury, a temporary illness, will often end in death, because it leaves the individual powerless against its enemies.   If an herbivorous animal is a little sick and has not fed well for a day or two, and the herd is then pursued by a beast of prey, our poor invalid inevitably falls a victim. So, in a carnivorous animal, the least deficiency of vigour prevents its capturing food, and it soon dies of starvation. There is, as a general rule, no mutual assistance between adults, which enables them to tide over a period of sickness.' Neither is there any division of labour; each must fulfil *all* the conditions of its existence, and, therefore, " natural selection " keeps all up to a pretty uniform standard.

But in man, as we now behold him, this is different. He is social and sympathetic. In the rudest tribes the sick are assisted, at least with food; less robust health and vigour than the average does not entail death. Neither does the want of perfect limbs, or other organs, produce the same effects as among animals. Some division of labour takes place ; the swiftest hunt, the less active fish, or gather fruits; food is, to some extent, exchanged or divided. The action of natural selection is therefore checked ; the weaker, the dwarfish, those of less active limbs, or less piercing eyesight, do not suffer the extreme penalty which falls upon animals so defective.

In proportion as these physical characteristics become of less importance, mental and moral qualities will have increasing influence on the well-being of the race. Capacity for acting in concert for protection, and for the acquisition of food and shelter; sympathy, which leads all in turn to assist each other; the sense of right, which checks depredations upon our fellows; the smaller development of the combative and destructive propensities; self-restraint in present appetites; and that intelligent foresight which prepares for the future, are all qualities, that from their earliest appearance must have been for the benefit of each community, and would, therefore, have become the subjects of " natural selection." For it is evident that such qualities would be for the well-being of man; would guard him against external enemies, against internal dissensions, and against the effects of inclement seasons and impending famine, more surely than could any merely physical modification. Tribes in which such mental and moral qualities were predominant, would therefore have an advantage in the struggle for existence over other tribes in which they were less developed, would live and maintain their numbers, while the others would decrease and finally succumb.

Again, when any slow changes of physical geography, or of climate, make it necessary for an animal to alter its food, its clothing, or its weapons, it can only do so by the occurrence of a corresponding change in its own bodily structure and internal organization. If a larger or more powerful beast is to be captured and devoured, as when a carnivorous animal which has hitherto preyed on antelopes is obliged from their decreasing numbers to attack buffaloes, it is only the strongest who can hold,—those with most powerful claws, and formidable canine teeth, that can struggle with and overcome such an animal. Natural selection immediately comes into play, and by its action these organs gradually become adapted to their new requirements. But man, under similar circumstances, does not require longer nails or teeth, greater bodily strength or swiftness. He makes sharper spears, or a better bow, or he constructs a cunning pitfall, or combines in a hunting party to circumvent his new prey. The capacities which enable him to do this are what he requires to be strengthened, and these will, therefore, be gradually modified by "natural selection," while the form and structure of his body will remain unchanged. So, when a glacial epoch comes on, some animals must acquire warmer fur, or a covering of fat, or else die of cold. Those best clothed by nature are, therefore, preserved by natural selection. Man, under the same circumstances, will make himself warmer clothing, and build better houses; and the necessity of doing this will react upon his mental organization and social condition—will advance them while his natural body remains naked as before.

When the accustomed food of some animal becomes scarce or totally fails, it can only exist by becoming adapted to a new kind of food, a food perhaps less nourishing and less digestible. "Natural selection" will now act upon the stomach and intestines, and all their individual variations will be taken advantage of, to modify the race into harmony with its new food. In many cases, however, it is probable that this cannot be done. The internal organs may not vary quick enough, and then the animal will decrease in numbers, and finally become extinct. But man guards himself from such accidents by superintending and guiding the operations of nature. He plants the seed of his most agreeable food, and thus procures a supply, independent of the accidents of varying seasons or natural extinction. He domesticates animals, which serve him either to capture food or for food itself, and thus, changes of any great extent in his teeth or digestive organs are rendered unnecessary. Man, too, has everywhere the use of fire, and by its means can render palatable a variety of animal and vegetable substances, which he could hardly otherwise make use of; and thus obtains for himself a supply of food far more varied and abundant than that which any animal can command.

Thus man, by the mere capacity of clothing himself, and making weapons and tools, has taken away from nature that power of slowly but permanently changing the external form and structure, in accordance with changes in the external world, which she exercises over all other animals. As the competing races by which they are surrounded, the climate, the vegetation, or the animals which serve them for food, are slowly changing, they must undergo a corresponding change in their structure, habits, and constitution, to keep them in harmony with the new conditions—to enable them to live and maintain their numbers. But man does this by means of his intellect alone, the variations of which enable him, with an unchanged body, still to keep in harmony with the changing universe.

There is one point, however, in which nature will still act upon him as it does on animals, and, to some extent, modify his external characters. Mr. Darwin has shown, that the colour of the skin is correlated with constitutional peculiarities both in vegetables and animals, so that liability to certain diseases or freedom from them is often accompanied by marked external characters. Now, there is every reason to believe that this has acted, and, to some extent, may still continue to act, on man. In localities where certain diseases are prevalent, those individuals of savage races which were subject to them would rapidly die off; while those who were constitutionally free from the disease would survive, and form the progenitors of a new race. These favoured individuals would probably be distinguished by peculiarities of *colour*, with which again peculiarities in the texture or the abundance of *hair* seem to be correlated, and thus may have been brought about those racial differences of colour, which seem to have no relation to mere temperature or other obvious peculiarities of climate.

From the time, therefore, when the social and sympathetic feelings came into active operation, and the intellectual and moral faculties became fairly developed, man would cease to be influenced by " natural selection " in his physical form and structure. As an animal he would remain almost stationary, the changes of the surrounding universe ceasing to produce in him that powerful modifying effect which they exercise over other parts of the organic world. But from the moment that the form of his body became stationary, his mind would become subject to those very influences from which his body had escaped; every slight variation in his mental and moral nature which should enable him better to guard against adverse circumstances, and combine for mutual comfort and protection, would be preserved and accumulated ; the better and higher specimens of our race would therefore increase and spread, the lower and more brutal would give way and successively die out, and that rapid advancement of mental organization would occur, which has raised the very lowest races of man so far above the brutes (although differing so little from some of them in physical structure), and, in conjunction with scarcely perceptible modifications of form, has developed the wonderful intellect of the European races.

Influence of external Nature in the development of the Human Mind.

But from the time when this mental and moral advance commenced, and man's physical character became fixed and almost immutable, a new series of causes would come into action, and take part in his mental growth. The diverse aspects of nature would now make themselves felt, and profoundly influence the character of the primitive man.

When the power that had hitherto modified the body had its action transferred to the mind, then races would advance and become improved, merely by the harsh discipline of a sterile soil and inclement seasons. Under their influence, a hardier, a more provident, and a more social race would be developed, than in those regions where the earth produces a perennial supply of vegetable food, and where neither foresight nor ingenuity are required to prepare for the rigours of winter. And is it not the fact that in all ages, and in every quarter of the globe, the inhabitants of temperate have been superior to those of hotter countries? All the great invasions and displacements of races have been from North to South, rather than the reverse; and we have no record of there ever having existed, any more than there exists to-day, a solitary instance of an indigenous inter-tropical civilization. The Mexican civilization and government came from the North, and, as well as the Peruvian, was established, not in the rich tropical plains, but on the lofty and sterile plateaux of the Andes. The religion and civilization of Ceylon were introduced from North India; the successive conquerors of the Indian peninsula came from the North-west; the northern Mongols conquered the more Southern Chinese; and it was the bold and adventurous tribes of the North that overran and infused new life into Southern Europe.

Extinction of Lower Races.

It is the same great law of " the preservation of favoured races in the struggle for life," which leads to the inevitable extinction of all those low and mentally undeveloped populations with which Europeans come in contact. The red Indian in North

America, and in Brazil; the Tasmanian, Australian, and New Zealander in the southern hemisphere, die out, not from any one special cause, but from the inevitable effects of an unequal mental and physical struggle. The intellectual and moral, as well as the physical, qualities of the European are superior; the same powers and capacities which have made him rise in a few centuries from the condition of the wandering savage with a scanty and stationary population, to his present state of culture and advancement, with a greater average longevity, a greater average strength, and a capacity of more rapid increase,— enable him when in contact with the savage man, to conquer in the struggle for existence, and to increase at his expense, just as the better adapted, increase at the expense of the less adapted varieties in the animal and vegetable kingdoms,—just as the weeds of Europe overrun North America and Australia, extinguishing native productions by the inherent vigour of their organization, and by their greater capacity for existence and multiplication.

The Origin of the Races of Man.

If these views are correct; if in proportion as man's social, moral, and intellectual faculties became developed, his physical structure would cease to be affected by the operation of "natural selection," we have a most important clue to the origin of races. For it will follow, that those great modifications of structure and of external form, which resulted in the development of man out of some lower type of animal, must have occurred before his intellect had raised him above the condition of the brutes, at a period when he was gregarious, but scarcely social, with a mind perceptive but not reflective, ere any sense of *right* or feelings of *sympathy* had been developed in him. He would be still subject, like the rest of the organic world, to the action of "natural selection," which would retain his physical form and constitution in harmony with the surrounding universe. He was probably at a very early period a dominant race, spreading widely over the warmer regions of the earth as it then existed, and in agreement with what we see in the case of other dominant species, gradually becoming modified in accordance with local conditions. As he ranged farther from his original home, and became exposed to greater extremes of climate, to greater changes of food, and had to contend with new enemies, organic and inorganic, slight useful variations in his constitution would be selected and rendered permanent, and would, on the principle of "correlation of growth," be accompanied by corresponding external physical changes. Thus might have arisen those striking characteristics and special modifications which still distinguish the chief races of mankind. The red, black, yellow, or blushing white skin; the straight, the curly, the woolly hair; the scanty or abundant beard; the straight or oblique eyes; the various forms of the pelvis, the cranium, and other parts of the skeleton.

But while these changes had been going on, his

mental development had, from some unknown cause, greatly advanced, and had now reached that condition in which it began powerfully to influence his whole existence, and would therefore become subject to the irresistible action of "natural selection." This action would quickly give the ascendency to mind: speech would probably now be first developed, leading to a still further advance of the mental faculties; and from that moment man, as regards the form and structure of most parts of his body, would remain almost stationary. The art of making weapons, division of labour, anticipation of the future, restraint of the appetites, moral, social, and sympathetic feelings, would now have a preponderating influence on his well being, and would therefore be that part of his nature on which "natural selection" would most powerfully act; and we should thus have explained that wonderful persistence of mere physical characteristics, which is the stumbling-block of those who advocate the unity of mankind.

We are now, therefore, enabled to harmonise the conflicting views of anthropologists on this subject. Man may have been, indeed I believe must have been, once a homogeneous race; but it was at a period of which we have as yet discovered no remains, at a period so remote in his history, that he had not yet acquired that wonderfully developed brain, the organ of the mind, which now, even in his lowest examples, raises him far above the highest brutes;—at a period when he had the form but hardly the nature of man, when

Y

he neither possessed human speech, nor those sympathetic and moral feelings which in a greater or less degree everywhere now distinguish the race. Just in proportion as these truly human faculties became developed in him, would his physical features become fixed and permanent, because the latter would be of less importance to his well being; he would be kept in harmony with the slowly changing universe around him, by an advance in mind, rather than by a change in body. If, therefore, we are of opinion that he was not really man till these higher faculties were fully developed, we may fairly assert that there were many originally distinct races of men; while, if we think that a being closely resembling us in form and structure, but with mental faculties scarcely raised above the brute, must still be considered to have been human, we are fully entitled to maintain the common origin of all mankind.

The Bearing of these Views on the Antiquity of Man.

These considerations, it will be seen, enable us to place the origin of man at a much more remote geological epoch than has yet been thought possible. He may even have lived in the Miocene or Eocene period, when not a single mammal was identical in form with any existing species. For, in the long series of ages during which these primeval animals were being slowly changed into the species which now inhabit the earth, the power which acted to modify them would only affect the mental organization of man. His brain alone would have increased in size and complexity, and his cranium have undergone corresponding changes of form, while the whole structure of lower animals was being changed. This will enable us to understand how the fossil crania of Denise and Engis agree so closely with existing forms, although they undoubtedly existed in company with large mammalia now extinct. The Neanderthal skull may be a specimen of one of the lowest races then existing, just as the Australians are the lowest of our modern epoch. We have no reason to suppose that mind and brain and skull modification, could go on quicker than that of the other parts of the organization; and we must therefore look back very far in the past, to find man in that early condition in which his mind was not sufficiently developed, to remove his body from the modifying influence of external conditions and the cumulative action of " natural selection." I believe, therefore, that there is no *à priori* reason against our finding the remains of man or his works in the tertiary deposits. The absence of all such remains in the European beds of this age has little weight, because, as we go further back in time, it is natural to suppose that man's distribution over the surface of the earth was less universal than at present.

Besides, Europe was in a great measure submerged during the tertiary epoch; and though its scattered islands may have been uninhabited by man, it by no means follows that he did not at the same time exist in warm or tropical continents. If geologists can point out to us the most extensive land in the warmer regions of the earth, which has not been submerged since Eocene or Miocene times, it is there that we may expect to find some traces of the very early progenitors of man. It is there that we may trace back the gradually decreasing brain of former races, till we come to a time when the body also begins materially to differ. Then we shall have reached the starting point of the human family. Before that period, he had not mind enough to preserve his body from change, and would, therefore, have been subject to the same comparatively rapid modifications of form as the other mammalia.

Their Bearing on the Dignity and Supremacy of Man.

If the views I have here endeavoured to sustain have any foundation, they give us a new argument for placing man apart, as not only the head and culminating point of the grand series of organic nature, but as in some degree a new and distinct order of being. From those infinitely remote ages, when the first rudiments of organic life appeared upon the earth, every plant, and every animal has been subject to one great law of physical change. As the earth has gone through its grand cycles of geological, climatal, and organic progress, every form of life has been subject to its irresistible action, and has been continually, but imperceptibly moulded into such new shapes as would preserve their harmony with the ever-changing universe. No living thing could escape this law of its being; none (except, perhaps, the simplest and most rudimentary organisms), could remain unchanged and live, amid the universal change around it.

At length, however, there came into existence a being in whom that subtle force we term *mind*, became of greater importance than his mere bodily structure. Though with a naked and unprotected body, *this* gave him clothing against the varying inclemencies of the seasons. Though unable to compete with the deer in swiftness, or with the wild bull in strength, *this* gave him weapons with which to capture or overcome both. Though less capable than most other animals of living on the herbs and the fruits that unaided nature supplies, this wonderful faculty taught him to govern and direct nature to his own benefit, and make her produce food for him, when and where he pleased. From the moment when the first skin was used as a covering, when the first rude spear was formed to assist in the chase, when fire was first used to cook his food, when the first seed was sown or shoot planted, a grand revolution was effected in nature, a revolution which in all the previous ages of the earth's history had had no parallel, for a being had arisen who was no longer necessarily subject to change with the changing universe—a being who was in some degree superior to nature, inasmuch as he knew how to control and regulate her action, and could keep himself in harmony with her, not by a change in body, but by an advance of mind.

Here, then, we see the true grandeur and dignity of man. On this view of his special attributes, we

may admit, that even those who claim for him a
position as an order, a class, or a sub-kingdom by
himself, have some show of reason on their side. He
is, indeed, a being apart, since he is not influenced
by the great laws which irresistibly modify all other
organic beings. Nay more ; this victory which he has
gained for himself, gives him a directing influence
over other existences. Man has not only escaped
" natural selection " himself, but he is actually able
to take away some of that power from nature which
before his appearance she universally exercised. We
can anticipate the time when the earth will produce
only cultivated plants and domestic animals ; when
man's selection shall have supplanted " natural selection ;" and when the ocean will be the only domain
in which that power can be exerted, which for countless cycles of ages ruled supreme over all the earth.

Their Bearing on the future Development of Man.

We now find ourselves enabled to answer those who
maintain, that if Mr. Darwin's theory of the Origin of
Species is true, man too must change in form, and become developed into some other animal as different from
his present self as he is from the Gorilla or the Chimpanzee ; and who speculate on what this form is likely
to be. But it is evident that such will not be the case ;
for no change of conditions is conceivable, which will
render any important alteration of his form and organization so universally useful and necessary to him, as
to give those possessing it always the best chance of

surviving, and thus lead to the development of a new species, genus, or higher group of man. On the other hand, we know that far greater changes of conditions and of his entire environment have been undergone by man, than any other highly organized animal could survive unchanged, and have been met by mental, not corporeal adaptation. The difference of habits, of food, clothing, weapons, and enemies, between savage and civilized man, is enormous. Difference in bodily form and structure there is practically none, except a slightly increased size of brain, corresponding to his higher mental development.

We have every reason to believe, then, that man may have existed and may continue to exist, through a series of geological periods which shall see all other forms of animal life again and again changed; while he himself remains unchanged, except in the two particulars already specified—the head and face, as immediately connected with the organ of the mind and as being the medium of expressing the most refined emotions of his nature,—and to a slight extent in colour, hair, and proportions, so far as they are correlated with constitutional resistance to disease.

Summary.

Briefly to recapitulate the argument;—in two distinct ways has man escaped the influence of those laws which have produced unceasing change in the animal world. 1. By his superior intellect he is enabled to provide himself with clothing and weapons, and by cultivating the soil to obtain a constant supply of congenial food. This renders it unnecessary for his body, like those of the lower animals, to be modified in accordance with changing conditions—to gain a warmer natural covering, to acquire more powerful teeth or claws, or to become adapted to obtain and digest new kinds of food, as circumstances may require. 2. By his superior sympathetic and moral feelings, he becomes fitted for the social state; he ceases to plunder the weak and helpless of his tribe; he shares the game which he has caught with less active or less fortunate hunters, or exchanges it for weapons which even the weak or the deformed can fashion; he saves the sick and wounded from death; and thus the power which leads to the rigid destruction of all animals who cannot in every respect help themselves, is prevented from acting on him.

This power is "natural selection;" and, as by no other means can it be shown, that individual variations can ever become accumulated and rendered permanent so as to form well-marked races, it follows that the differences which now separate mankind from other animals, must have been produced before he became possessed of a human intellect or human sympathies. This view also renders possible, or even requires, the existence of man at a comparatively remote geological epoch. For, during the long periods in which other animals have been undergoing modification in their whole structure, to such an amount as to constitute distinct genera and families, man's *body* will have remained generically, or even specifically, the same, while his *head* and *brain* alone will have undergone modification equal to theirs. We can thus understand how it is that, judging from the head and brain, Professor Owen places man in a distinct sub-class of mammalia, while as regards the bony structure of his body, there is the closest anatomical resemblance to the anthropoid apes, "every tooth, every bone, strictly homologous—which makes the determination of the difference between *Homo* and *Pithecus* the anatomist's difficulty." The present theory fully recognises and accounts for these facts ; and we may perhaps claim as corroborative of its truth, that it neither requires us to depreciate the intellectual chasm which separates man from the apes, nor refuses full recognition of the striking resemblances to them, which exist in other parts of his structure.

Conclusion.

In concluding this brief sketch of a great subject, I would point out its bearing upon the future of the human race. If my conclusions are just, it must inevitably follow that the higher—the more intellectual and moral—must displace the lower and more degraded races ; and the power of " natural selection," still acting on his mental organization, must ever lead to the more perfect adaptation of man's higher faculties to the conditions of surrounding nature, and to the exigencies of the social state. While his external form will probably ever remain unchanged, except in

the development of that perfect beauty which results
from a healthy and well organized body, refined and
ennobled by the highest intellectual faculties and sympathetic emotions, his mental constitution may continue to advance and improve, till the world is again
inhabited by a single nearly homogeneous race, no
individual of which will be inferior to the noblest
specimens of existing humanity.

Our progress towards such a result is very slow, but
it still seems to be a progress. We are just now living
at an abnormal period of the world's history, owing to
the marvellous developments and vast practical results
of science, having been given to societies too low
morally and intellectually, to know how to make the
best use of them, and to whom they have consequently
been curses as well as blessings. Among civilized nations at the present day, it does not seem possible for
natural selection to act in any way, so as to secure the
permanent advancement of morality and intelligence;
for it is indisputably the mediocre, if not the low, both
as regards morality and intelligence, who succeed best
in life and multiply fastest. Yet there is undoubtedly
an advance—on the whole a steady and a permanent
one—both in the influence on public opinion of a high
morality, and in the general desire for intellectual elevation; and as I cannot impute this in any way to
"survival of the fittest," I am forced to conclude that
it is due, to the inherent progressive power of those
glorious qualities which raise us so immeasurably above
our fellow animals, and at the same time afford us the

surest proof that there are other and higher existences than ourselves, from whom these qualities may have been derived, and towards whom we may be ever tending.

X.

THE LIMITS OF NATURAL SELECTION AS APPLIED TO MAN.

THROUGHOUT this volume I have endeavoured to show, that the known laws of variation, multiplication, and heredity, resulting in a "struggle for existence" and the "survival of the fittest," have probably sufficed to produce all the varieties of structure, all the wonderful adaptations, all the beauty of form and of colour, that we see in the animal and vegetable kingdoms. To the best of my ability I have answered the most obvious and the most often repeated objections to this theory, and have, I hope, added to its general strength, by showing how colour—one of the strongholds of the advocates of special creation—may be, in almost all its modifications, accounted for by the combined influence of sexual selection and the need of protection. I have also endeavoured to show, how the same power which has modified animals has acted on man; and have, I believe, proved that, as soon as the human intellect became developed above a certain low stage, man's body would cease to be materially affected by natural selection, because the development of his mental faculties would render important modifications of its form and structure unnecessary. It will, therefore, probably excite some surprise among my readers, to find that I do not consider that all nature can be explained on the principles of which I am so ardent an advocate; and that I am now myself going to state objections, and to place limits, to the power of "natural selection." I believe, however, that there are such limits; and that just as surely as we can trace the action of natural laws in the development of organic forms, and can clearly conceive that fuller knowledge would enable us to follow step by step the whole process of that development, so surely can we trace the action of some unknown higher law, beyond and independent of all those laws of which we have any knowledge. We can trace this action more or less distinctly in many phenomena, the two most important of which are—the origin of sensation or consciousness, and the development of man from the lower animals. I shall first consider the latter difficulty as more immediately connected with the subjects discussed in this volume.

What Natural Selection can Not do.

In considering the question of the development of man by known natural laws, we must ever bear in mind the first principle of "natural selection," no less than of the general theory of evolution, that all changes of form or structure, all increase in the size of an organ or in its complexity, all greater specialization or physiological division of labour, can only be brought about, in as much as it is for the good of the being so modified. Mr. Darwin himself has taken care to impress upon us, that "natural selection" has no power to produce absolute perfection but only relative perfection, no power to advance any being much beyond his fellow beings, but only just so much beyond them as to enable it to survive them in the struggle for existence.   Still less has it any power to produce modifications which are in any degree injurious to its possessor, and Mr. Darwin frequently uses the strong expression, that a single case of this kind would be fatal to his theory.   If, therefore, we find in man any characters, which all the evidence we can obtain goes to show would have been actually injurious to him on their first appearance, they could not possibly have been produced by natural selection.   Neither could any specially developed organ have been so produced if it had been merely useless to him, or if its use were not proportionate to its degree of development.   Such cases as these would prove, that some other law, or some other power, than "natural selection" had been at work.   But if, further, we could see that these very modifications, though hurtful or useless at the time when they first appeared, became in the highest degree useful at a much later period, and are now essential to the full moral and intellectual development of human nature, we should then infer the action of mind, foreseeing the future and preparing for it, just as surely as we do, when we see the breeder set himself to work with the determination to produce a definite improvement in some cultivated plant or domestic animal.   I would further remark that this enquiry is as thoroughly scientific and legitimate as that into the origin of species itself.   It is an attempt to solve the inverse problem, to deduce the existence of a new power of a definite character, in order to account for facts which according to the theory of natural selection ought not to happen.   Such problems are well known to science, and the search after their solution has often led to the most brilliant results.   In the case of man, there are facts of the nature above alluded to, and in calling attention to them, and in inferring a cause for them, I believe that I am as strictly within the bounds of scientific investigation as I have been in any other portion of my work.

The Brain of the Savage shown to be Larger than he Needs it to be.

Size of Brain an important Element of Mental Power.—The brain is universally admitted to be the organ of the mind; and it is almost as universally admitted, that size of brain is one of the most important of the elements which determine mental power or capacity.   There seems to be no doubt that brains differ considerably in quality, as indicated by greater or less complexity of the convolutions, quantity of grey matter, and perhaps unknown peculiarities of organization ; but this difference of quality seems merely to increase or diminish the influence of quantity, not to neutralize it.   Thus, all the most eminent modern writers see an intimate connection between the di- minished size of the brain in the lower races of mankind, and their intellectual inferiority. The collections of Dr. J. B. Davis and Dr. Morton give the following as the average internal capacity of the cranium in the chief races:—Teutonic family, 94 cubic inches; Esquimaux, 91 cubic inches; Negroes, 85 cubic inches; Australians and Tasmanians, 82 cubic inches; Bushmen, 77 cubic inches. These last numbers, however, are deduced from comparatively few specimens, and may be below the average, just as a small number of Finns and Cossacks give 98 cubic inches, or considerably more than that of the German races. It is evident, therefore, that the absolute bulk of the brain is not necessarily much less in savage than in civilised man, for Esquimaux skulls are known with a capacity of 113 inches, or hardly less than the largest among Europeans. But what is still more extraordinary, the few remains yet known of pre-historic man do not indicate any material diminution in the size of the brain case. A Swiss skull of the stone age, found in the lake dwelling of Meilen, corresponded exactly to that of a Swiss youth of the present day. The celebrated Neanderthal skull had a larger circumference than the average, and its capacity, indicating actual mass of brain, is estimated to have been not less than 75 cubic inches, or nearly the average of existing Australian crania. The Engis skull, perhaps the oldest known, and which, according to Sir John Lubbock, "there seems no doubt was really contemporary with the mammoth and the cave bear," is yet, according to Professor Huxley, "a fair average skull, which might have belonged to a philosopher, or might have contained the thoughtless brains of a savage." Of the cave men of Les Eyzies, who were undoubtedly contemporary with the reindeer in the South of France, Professor Paul Broca says (in a paper read before the Congress of Pre-historic Archæology in 1868)— "The great capacity of the brain, the development of the frontal region, the fine elliptical form of the anterior part of the profile of the skull, are incontestible characteristics of superiority, such as we are accustomed to meet with in civilised races ; " yet the great breadth of the face, the enormous development of the ascending ramus of the lower jaw, the extent and roughness of the surfaces for the attachment of the muscles, especially of the masticators, and the extraordinary development of the ridge of the femur, indicate enormous muscular power, and the habits of a savage and brutal race.

These facts might almost make us doubt whether the size of the brain is in any direct way an index of mental power, had we not the most conclusive evidence that it is so, in the fact that, whenever an adult male European has a skull less than nineteen inches in circumference, or has less than sixty-five cubic inches of brain, he is invariably idiotic. When we join with this the equally undisputed fact, that great men—those who combine acute perception with great reflective power, strong passions, and general energy of character, such as Napoleon, Cuvier, and O'Connell, have always heads far above the average size, we must feel satisfied that

z

volume of brain is one, and perhaps the most important, measure of intellect ; and this being the case, we
cannot fail to be struck with the apparent anomaly,
that many of the lowest savages should have as much
brains as average Europeans. The idea is suggested
of a surplusage of power ; of an instrument beyond the
needs of its possessor.

*Comparison of the Brains of Man and of Anthropoid
Apes.*—In order to discover if there is any foundation
for this notion, let us compare the brain of man with
that of animals. The adult male Orang-utan is quite as
bulky as a small sized man, while the Gorilla is considerably above the average size of man, as estimated by
bulk and weight ; yet the former has a brain of only
28 cubic inches, the latter, one of 30, or, in the largest
specimen yet known, of $34\frac{1}{2}$ cubic inches. We have
seen that the average cranial capacity of the lowest
savages is probably not less than *five-sixths* of that of
the highest civilized races, while the brain of the
anthropoid apes scarcely amounts to *one-third* of that
of man, in both cases taking the average ; or the
proportions may be more clearly represented by the
following figures—anthropoid apes, 10 ; savages, 26 ;
civilized man, 32. But do these figures at all approximately represent the relative intellect of the three
groups ? Is the savage really no further removed from
the philosopher, and so much removed from the ape,
as these figures would indicate ? In considering this
question, we must not forget that the heads of savages
vary in size, almost as much as those of civilized

Europeans. Thus, while the largest Teutonic skull in Dr. Davis' collection is 112·4 cubic inches, there is an Araucanian of 115·5, an Esquimaux of 113·1, a Marquesan of 110·6, a Negro of 105·8, and even an Australian of 104·5 cubic inches. We may, therefore, fairly compare the savage with the highest European on the one side, and with the Orang, Chimpanzee, or Gorilla, on the other, and see whether there is any relative proportion between brain and intellect.

Range of intellectual power in Man.—First, let us consider what this wonderful instrument, the brain, is capable of in its higher developments. In Mr. Galton's interesting work on " Hereditary Genius," he remarks on the enormous difference between the intellectual power and grasp of the well-trained mathematician or man of science, and the average Englishman. The number of marks obtained by high wranglers, is often more than thirty times as great as that of the men at the bottom of the honour list, who are still of fair mathematical ability ; and it is the opinion of skilled examiners, that even this does not represent the full difference of intellectual power. If, now, we descend to those savage tribes who only count to three or five, and who find it impossible to comprehend the addition of two and three without having the objects actually before them, we feel that the chasm between them and the good mathematician is so vast, that a thousand to one will probably not fully express it. Yet we know that the mass of brain might be nearly the same in both, or might not differ in a greater proportion than as 5 to 6 ; whence we may fairly infer that the savage possesses a brain capable, if cultivated and developed, of performing work of a kind and degree far beyond what he ever requires it to do.

Again, let us consider the power of the higher or even the average civilized man, of forming abstract ideas, and carrying on more or less complex trains of reasoning. Our languages are full of terms to express abstract conceptions. Our business and our pleasures involve the continual foresight of many contingencies. Our law, our government, and our science, continually require us to reason through a variety of complicated phenomena to the expected result. Even our games, such as chess, compel us to exercise all these faculties in a remarkable degree. Compare this with the savage languages, which contain no words for abstract conceptions ; the utter want of foresight of the savage man beyond his simplest necessities ; his inability to combine, or to compare, or to reason on any general subject that does not immediately appeal to his senses. So, in his moral and æsthetic faculties, the savage has none of those wide sympathies with all nature, those conceptions of the infinite, of the good, of the sublime and beautiful, which are so largely developed in civilized man. Any considerable development of these would, in fact, be useless or even hurtful to him, since they would to some extent interfere with the supremacy of those perceptive and animal faculties on which his very existence often depends, in the severe struggle he has to carry on against nature and his fellow-man. Yet the rudiments of all these powers and feelings undoubtedly exist in him, since one or other of them frequently manifest themselves in exceptional cases, or when some special circumstances call them forth. Some tribes, such as the Santals, are remarkable for as pure a love of truth as the most moral among civilized men. The Hindoo and the Polynesian have a high artistic feeling, the first traces of which are clearly visible in the rude drawings of the palæolithic men who were the contemporaries in France of the Reindeer and the Mammoth. Instances of unselfish love, of true gratitude, and of deep religious feeling, sometimes occur among most savage races.

On the whole, then, we may conclude, that the general moral and intellectual development of the savage, is not less removed from that of civilized man than has been shown to be the case in the one department of mathematics ; and from the fact that all the moral and intellectual faculties do occasionally manifest themselves, we may fairly conclude that they are always latent, and that the large brain of the savage man is much beyond his actual requirements in the savage state.

Intellect of Savages and of Animals compared.—Let us now compare the intellectual wants of the savage, and the actual amount of intellect he exhibits, with those of the higher animals. Such races as the Andaman Islanders, the Australians, and the Tasmanians, the Digger Indians of North America, or the natives of Fuegia, pass their lives so as to require the exercise of few faculties not possessed in an equal degree by many animals. In the mode of capture of game or fish, they by no means surpass the ingenuity or forethought of the jaguar, who drops saliva into the water, and seizes the fish as they come to eat it; or of wolves and jackals, who hunt in packs; or of the fox, who buries his surplus food till he requires it. The sentinels placed by antelopes and by monkeys, and the various modes of building adopted by field mice and beavers, as well as the sleeping place of the orang-utan, and the tree-shelter of some of the African anthropoid apes, may well be compared with the amount of care and forethought bestowed by many savages in similar circumstances. His possession of free and perfect hands, not required for locomotion, enable man to form and use weapons and implements which are beyond the physical powers of brutes; but having done this, he certainly does not exhibit more mind in using them than do many lower animals. What is there in the life of the savage, but the satisfy-ing of the cravings of appetite in the simplest and easiest way? What thoughts, ideas, or actions are there, that raise him many grades above the elephant or the ape? Yet he possesses, as we have seen, a brain vastly superior to theirs in size and complexity; and this brain gives him, in an undeveloped state, faculties which he never requires to use. And if this is true of existing savages, how much more true must it have been of the men whose sole weapons were rudely chipped flints, and some of whom, we may fairly conclude, were lower than any existing race; while the only evidence yet in our possession shows them to have had brains fully as capacious as those of the average of the lower savage races.

We see, then, that whether we compare the savage with the higher developments of man, or with the brutes around him, we are alike driven to the conclusion that in his large and well-developed brain he possesses an organ quite disproportionate to his actual requirements—an organ that seems prepared in advance, only to be fully utilized as he progresses in civilization. A brain slightly larger than that of the gorilla would, according to the evidence before us, fully have sufficed for the limited mental development of the savage; and we must therefore admit, that the large brain he actually possesses could never have been solely developed by any of those laws of evolution, whose essence is, that they lead to a degree of organization exactly proportionate to the wants of each species, never beyond those wants—that no preparation can be made for the future development of the race—that one part of the body can never increase in size or complexity, except in strict co-ordination to the pressing wants of the whole. The brain of prehistoric and of savage man seems to me to prove the existence of some power, distinct from that which has guided the development of the lower animals through their ever-varying forms of being.

The Use of the Hairy Covering of Mammalia.

Let us now consider another point in man's organization, the bearing of which has been almost entirely
overlooked by writers on both sides of this question.
One of the most general external characters of the
terrestrial mammalia is the hairy covering of the body,
which, whenever the skin is flexible, soft, and sensitive,
forms a natural protection against the severities of climate, and particularly against rain.   That this is its
most important function, is well shown by the manner
in which the hairs are disposed so as to carry off the
water, by being invariably directed downwards from
the most elevated parts of the body. Thus, on the under
surface the hair is always less plentiful, and, in many
cases, the belly is almost bare. The hair lies downwards, on the limbs of all walking mammals, from the
shoulder to the toes, but in the orang-utan it is directed
from the shoulder to the elbow, and again from the
wrist to the elbow, in a reverse direction. This corresponds to the habits of the animal, which, when resting,
holds its long arms upwards over its head, or clasping
a branch above it, so that the rain would flow down
both the arm and fore-arm to the long hair which meets
at the elbow. In accordance with this principle, the
hair is always longer or more dense along the spine
or middle of the back from the nape to the tail, often
rising into a crest of hair or bristles on the ridge of the
back. This character prevails through the entire series
of the mammalia, from the marsupials to the quadru-

mana, and by this long persistence it must have acquired such a powerful hereditary tendency, that we should expect it to reappear continually even after it had been abolished by ages of the most rigid selection ; and we may feel sure that it never could have been completely abolished under the law of natural selection, unless it had become so positively injurious as to lead to the almost invariable extinction of individuals possessing it.

The constant absence of Hair from certain parts of Man's Body a remarkable Phenomenon.

In man the hairy covering of the body has almost totally disappeared, and, what is very remarkable, it has disappeared more completely from the back than from any other part of the body. Bearded and beardless races alike have the back smooth, and even when a considerable quantity of hair appears on the limbs and breast, the back, and especially the spinal region, is absolutely free, thus completely reversing the characteristics of all other mammalia. The Ainos of the Kurile Islands and Japan are said to be a hairy race; but Mr. Bickmore, who saw some of them, and described them in a paper read before the Ethnological Society, gives no details as to where the hair was most abundant, merely stating generally, that " their chief peculiarity is their great abundance of hair, not only on the head and face, but over the whole body." This might very well be said of any man who had hairy limbs and breast, unless it was specially stated that his back was

hairy, which is not done in this case. The hairy family
in Birmah have, indeed, hair on the back rather longer
than on the breast, thus reproducing the true mammalian character, but they have still longer hair on the
face, forehead, and inside the ears, which is quite abnormal ; and the fact that their teeth are all very imperfect, shows that this is a case of monstrosity rather
than one of true reversion to the ancestral type of man
before he lost his hairy covering.

Savage Man feels the Want of this Hairy Covering.

We must now enquire if we have any evidence to
show, or any reason to believe, that a hairy covering to
the back would be in any degree hurtful to savage
man, or to man in any stage of his progress from his
lower animal form ; and if it were merely useless, could
it have been so entirely and completely removed as not
to be continually reappearing in mixed races? Let
us look to savage man for some light on these points.
One of the most common habits of savages is to use
some covering for the back and shoulders, even when
they have none on any other part of the body. The
early voyagers observed with surprise, that the Tasmanians, both men and women, wore the kangaroo-
skin, which was their only covering, not from any
feeling of modesty, but over the shoulders to keep the
back dry and warm. A cloth over the shoulders was
also the national dress of the Maories. The Patago-
nians wear a cloak or mantle over the shoulders, and
the Fuegians often wear a small piece of skin on the

back, laced on, and shifted from side to side as the wind blows. The Hottentots also wore a somewhat similar skin over the back, which they never removed, and in which they were buried. Even in the tropics most savages take precautions to keep their backs dry. The natives of Timor use the leaf of a fan palm, carefully stitched up and folded, which they always carry with them, and which, held over the back, forms an admirable protection from the rain. Almost all the Malay races, as well as the Indians of South America, make great palm-leaf hats, four feet or more across, which they use during their canoe voyages to protect their bodies from heavy showers of rain; and they use smaller hats of the same kind when travelling by land.

We find, then, that so far from there being any reason to believe that a hairy covering to the back could have been hurtful or even useless to pre-historic man, the habits of modern savages indicate exactly the opposite view, as they evidently feel the want of it, and are obliged to provide substitutes of various kinds. The perfectly erect posture of man, may be supposed to have something to do with the disappearance of the hair from his body, while it remains on his head; but when walking, exposed to rain and wind, a man naturally stoops forwards, and thus exposes his back; and the undoubted fact, that most savages feel the effects of cold and wet most severely in that part of the body, sufficiently demonstrates that the hair could not have ceased to grow there merely because it was useless, even if it were likely that a character so long persistent in the entire order of mammalia, could have so completely disappeared, under the influence of so weak a selective power as a diminished usefulness.

Man's Naked Skin could not have been produced by Natural Selection.

It seems to me, then, to be absolutely certain, that " Natural Selection " could not have produced man's hairless body by the accumulation of variations from a hairy ancestor. The evidence all goes to show that such variations could not have been useful, but must, on the contrary, have been to some extent hurtful. If even, owing to an unknown correlation with other hurtful qualities, it had been abolished in the ancestral tropical man, we cannot conceive that, as man spread into colder climates, it should not have returned under the powerful influence of reversion to such a long persistent ancestral type. But the very foundation of such a supposition as this is untenable ; for we cannot suppose that a character which, like hairiness, exists throughout the whole of the mammalia, can have become, in one form only, so constantly correlated with an injurious character, as to lead to its permanent suppression—a suppression so complete and effectual that it never, or scarcely ever, reappears in mongrels of the most widely different races of man.

Two characters could hardly be wider apart, than the size and development of man's brain, and the distribution of hair upon the surface of his body ; yet they both lead us to the same conclusion—that some other power than Natural Selection has been engaged in his production.

Feet and Hands of Man, considered as Difficulties on the Theory of Natural Selection.

There are a few other physical characteristics of man, that may just be mentioned as offering similar difficulties, though I do not attach the same importance to them as to those I have already dwelt on. The specialization and perfection of the hands and feet of man seems difficult to account for. Throughout the whole of the quadrumana the foot is prehensile; and a very rigid selection must therefore have been needed to bring about that arrangement of the bones and muscles, which has converted the thumb into a great toe, so completely, that the power of opposability is totally lost in every race, whatever some travellers may vaguely assert to the contrary. It is difficult to see why the prehensile power should have been taken away. It must certainly have been useful in climbing, and the case of the baboons shows that it is quite compatible with terrestrial locomotion. It may not be compatible with perfectly easy erect locomotion; but, then, how can we conceive that early man, *as an animal,* gained anything by purely erect locomotion? Again, the hand of man contains latent capacities and powers which are unused by savages, and must have been even less used by palæolithic man and his still ruder predecessors. It has all the appearance of an organ prepared for the use of civilized man, and one which was required to render civilization possible. Apes make little use of their separate fingers and opposable thumbs. They grasp objects rudely and clumsily, and look as if a much less specialized extremity would have served their purpose as well. I do not lay much stress on this, but, if it be proved that some intelligent power has guided or determined the development of man, then we may see indications of that power, in facts which, by themselves, would not serve to prove its existence.

The voice of man.—The same remark will apply to another peculiarly human character, the wonderful power, range, flexibility, and sweetness, of the musical sounds producible by the human larynx, especially in the female sex. The habits of savages give no indication of how this faculty could have been developed by natural selection; because it is never required or used by them. The singing of savages is a more or less monotonous howling, and the females seldom sing at all. Savages certainly never choose their wives for fine voices, but for rude health, and strength, and physical beauty. Sexual selection could not therefore have developed this wonderful power, which only comes into play among civilized people. It seems as if the organ had been prepared in anticipation of the future progress of man, since it contains latent capacities which are useless to him in his earlier condition. The delicate correlations of structure that give it such marvellous powers, could not therefore have been acquired by means of natural selection.

The Origin of some of Man's Mental Faculties, by the preservation of Useful Variations, not possible.

Turning to the mind of man, we meet with many difficulties in attempting to understand, how those mental faculties, which are especially human, could have been acquired by the preservation of useful variations. At first sight, it would seem that such feelings as those of abstract justice and benevolence could never have been so acquired, because they are incompatible with the law of the strongest, which is the essence of natural selection. But this is, I think, an erroneous view, because we must look, not to individuals but to societies; and justice and benevolence, exercised towards members of the same tribe, would certainly tend to strengthen that tribe, and give it a superiority over another in which the right of the strongest prevailed, and where consequently the weak and the sickly were left to perish, and the few strong ruthlessly destroyed the many who were weaker.

But there is another class of human faculties that do not regard our fellow men, and which cannot, therefore, be thus accounted for. Such are the capacity to form ideal conceptions of space and time, of eternity and infinity—the capacity for intense artistic feelings of pleasure, in form, colour, and composition—and for those abstract notions of form and number which render geometry and arithmetic possible. How were all or any of these faculties first developed, when they could have been of no possible use to man in his early stages of barbarism ? How could " natural selection," or survival of the fittest in the struggle for existence, at all favour the development of mental powers so entirely removed from the material necessities of savage men, and which even now, with our comparatively high civilization, are, in their farthest developments, in advance of the age, and appear to have relation rather to the future of the race than to its actual status?

Difficulty as to the Origin of the Moral Sense.

Exactly the same difficulty arises, when we endeavour to account for the development of the moral sense or conscience in savage man ; for although the *practice* of benevolence, honesty, or truth, may have been useful to the tribe possessing these virtues, that does not at all account for the peculiar *sanctity*, attached to actions which each tribe considers right and moral, as contrasted with the very different feelings with which they regard what is merely *useful.* The utilitarian hypothesis (which is the theory of natural selection applied to the mind) seems inadequate to account for the development of the moral sense. This subject has been recently much discussed, and I will here only give one example to illustrate my argument. The utilitarian sanction for truthfulness is by no means very powerful or universal. Few laws enforce it. No very severe reprobation follows untruthfulness. In all ages and countries, falsehood has been thought allowable in love, and laudable in war; while, at the present day, it is held to be venial by the majority of mankind, in trade, commerce, and speculation. A certain amount of untruthfulness is a necessary part of politeness in the east and west alike, while even severe moralists have held a lie justifiable, to elude an enemy or prevent a crime. Such being the difficulties with which this virtue has had to struggle, with so many exceptions to its practice, with so many instances in which it brought ruin or death to its too ardent devotee, how can we believe that considerations of utility could ever invest it with the mysterious sanctity of the highest virtue,—could ever induce men to value truth for its own sake, and practice it regardless of consequences?

Yet, it is a fact, that such a mystical sense of wrong does attach to untruthfulness, not only among the higher classes of civilized people, but among whole tribes of utter savages. Sir Walter Elliott tells us (in his paper "On the Characteristics of the Population of Central and Southern India," published in the Journal of the Ethnological Society of London, vol. i., p. 107) that the Kurubars and Santals, barbarous hill-tribes of Central India, are noted for veracity. It is a common saying that "a Kurubar *always* speaks the truth;" and Major Jervis says, "the Santals are the most truthful men I ever met with." As a remarkable instance of this quality the following fact is given. A number of prisoners, taken during the

Santal insurrection, were allowed to go free on parole, to work at a certain spot for wages. After some time cholera attacked them and they were obliged to leave, but every man of them returned and gave up his earnings to the guard. Two hundred savages with money in their girdles, walked thirty miles back to prison rather than break their word! My own experience among savages has furnished me with similar, although less severely tested, instances; and we cannot avoid asking, how is it, that in these few cases " experiences of utility " have left such an overwhelming impression, while in so many others they have left none? The experiences of savage men as regards the utility of truth, must, in the long run, be pretty nearly equal. How is it, then, that in some cases the result is a sanctity which overrides all considerations of personal advantage, while in others there is hardly a rudiment of such a feeling?

The intuitional theory, which I am now advocating, explains this by the supposition, that there is a feeling— a sense of right and wrong—in our nature, antecedent to and independent of experiences of utility. Where free play is allowed to the relations between man and man, this feeling attaches itself to those acts of universal utility or self-sacrifice, which are the products of our affections and sympathies, and which we term moral; while it may be, and often is, perverted, to give the same sanction to acts of narrow and conventional utility which are really immoral,—as when the Hindoo will tell a lie, but will sooner starve than eat unclean food; and looks upon the marriage of adult females as gross immorality.

The strength of the moral feeling will depend upon individual or racial constitution, and on education and habit;—the acts to which its sanctions are applied, will depend upon how far the simple feelings and affections of our nature, have been modified by custom, by law, or by religion.

It is difficult to conceive that such an intense and mystical feeling of right and wrong, (so intense as to overcome all ideas of personal advantage or utility), could have been developed out of accumulated ancestral experiences of utility; and still more difficult to understand, how feelings developed by one set of utilities, could be transferred to acts of which the utility was partial, imaginary, or altogether absent. But if a moral sense is an essential part of our nature, it is easy to see, that its sanction may often be given to acts which are useless or immoral; just as the natural appetite for drink, is perverted by the drunkard into the means of his destruction.

Summary of the Argument as to the Insufficiency of Natural Selection to account for the Development of Man.

Briefly to resume my argument—I have shown that the brain of the lowest savages, and, as far as we yet know, of the pre-historic races, is little inferior in size to that of the highest types of man, and immensely superior to that of the higher animals; while it is

2 A 2

universally admitted that quantity of brain is one of the most important, and probably the most essential, of the elements which determine mental power. Yet the mental requirements of savages, and the faculties actually exercised by them, are very little above those of animals. The higher feelings of pure morality and refined emotion, and the power of abstract reasoning and ideal conception, are useless to them, are rarely if ever manifested, and have no important relations to their habits, wants, desires, or well-being. They possess a mental organ beyond their needs. Natural Selection could only have endowed savage man with a brain a little superior to that of an ape, whereas he actually possesses one very little inferior to that of a philosopher.

The soft, naked, sensitive skin of man, entirely free from that hairy covering which is so universal among other mammalia, cannot be explained on the theory of natural selection. The habits of savages show that they feel the want of this covering, which is most completely absent in man exactly where it is thickest in other animals. We have no reason whatever to believe, that it could have been hurtful, or even useless to primitive man; and, under these circumstances, its complete abolition, shown by its never reverting in mixed breeds, is a demonstration of the agency of some other power than the law of the survival of the fittest, in the development of man from the lower animals.

Other characters show difficulties of a similar kind, though not perhaps in an equal degree. The structure of the human foot and hand seem unnecessarily perfect for the needs of savage man, in whom they are as completely and as humanly developed as in the highest races. The structure of the human larynx, giving the power of speech and of producing musical sounds, and especially its extreme development in the female sex, are shown to be beyond the needs of savages, and from their known habits, impossible to have been acquired either by sexual selection, or by survival of the fittest.

The mind of man offers arguments in the same direction, hardly less strong than those derived from his bodily structure. A number of his mental faculties have no relation to his fellow men, or to his material progress. The power of conceiving eternity and infinity, and all those purely abstract notions of form, number, and harmony, which play so large a part in the life of civilised races, are entirely outside of the world of thought of the savage, and have no influence on his individual existence or on that of his tribe. They could not, therefore, have been developed by any preservation of useful forms of thought; yet we find occasional traces of them amidst a low civilization, and at a time when they could have had no practical effect on the success of the individual, the family, or the race; and the development of a moral sense or conscience by similar means is equally inconceivable.

But, on the other hand, we find that every one of these characteristics is necessary for the full development of human nature. The rapid progress of civilization under favourable conditions, would not be possible, were not the organ of the mind of man prepared in advance, fully developed as regards size, structure, and proportions, and only needing a few generations of use and habit to co-ordinate its complex functions. The naked and sensitive skin, by necessitating clothing and houses, would lead to the more rapid development of man's inventive and constructive faculties; and, by leading to a more refined feeling of personal modesty, may have influenced, to a considerable extent, his moral nature. The erect form of man, by freeing the hands from all locomotive uses, has been necessary for his intellectual advancement; and the extreme perfection of his hands, has alone rendered possible that excellence in all the arts of civilization which raises him so far above the savage, and is perhaps but the forerunner of a higher intellectual and moral advancement. The perfection of his vocal organs has first led to the formation of articulate speech, and then to the development of those exquisitely toned sounds, which are only appreciated by the higher races, and which are probably destined for more elevated uses and more refined enjoyment, in a higher condition than we have yet attained to. So, those faculties which enable us to transcend time and space, and to realize the wonderful conceptions of mathematics and philosophy, or which give us an intense yearning for abstract truth, (all of which were occasionally manifested at such an early period of human history as to be far in advance of any of the few practical applications which have since grown out of them), are evidently essential to the perfect development of man as a spiritual being, but are utterly inconceivable as having been produced through the action of a law which looks only, and can look only, to the immediate material welfare of the individual or the race.

The inference I would draw from this class of phenomena is, that a superior intelligence has guided the development of man in a definite direction, and for a special purpose, just as man guides the development of many animal and vegetable forms. The laws of evolution alone would, perhaps, never have produced a grain so well adapted to man's use as wheat and maize; such fruits as the seedless banana and bread-fruit; or such animals as the Guernsey milch cow, or the London dray-horse. Yet these so closely resemble the unaided productions of nature, that we may well imagine a being who had mastered the laws of development of organic forms through past ages, refusing to believe that any new power had been concerned in their production, and scornfully rejecting the theory (as my theory will be rejected by many who agree with me on other points), that in these few cases a controlling intelligence had directed the action of the laws of variation, multiplication, and survival, for his own purposes. We know, however, that this has been done; and we must therefore admit the possibility that, if we are not the highest intelligences in the universe, some higher intelligence may have directed the process by which the human race was developed, by means of more subtle agencies than we are acquainted with. At the same time I must confess, that this theory has the disadvantage of requiring the intervention of some distinct individual intelligence, to aid in the production of what we can hardly avoid considering as the ultimate aim and outcome of all organized existence—intellectual, ever-advancing, spiritual man. It therefore implies, that the great laws which govern the material universe were insufficient for his production, unless we consider (as we may fairly do) that the controlling action of such higher intelligences is a necessary part of those laws, just as the action of all surrounding organisms is one of the agencies in organic development. But even if my particular view should not be the true one, the difficulties I have put forward remain, and I think prove, that some more general and more fundamental law underlies that of " natural selection." The law of " unconscious intelligence" pervading all organic nature, put forth by Dr. Laycock and adopted by Mr. Murphy, is such a law ; but to my mind it has the double disadvantage of being both unintelligible and incapable of any kind of proof. It is more probable, that the true law lies too deep for us to discover it ; but there seems to me, to be ample indications that such a law does exist, and is probably connected with the absolute origin of life and organization. (*Note A.*)

The Origin of Consciousness.

The question of the origin of sensation and of thought can be but briefly discussed in this place, since it is a subject wide enough to require a separate volume for its proper treatment. No physiologist or philosopher has yet ventured to propound an intelligible theory, of how sensation may possibly be a product of organization; while many have declared the passage from matter to mind to be inconceivable. In his presidential address to the Physical Section of the British Association at Norwich, in 1868, Professor Tyndall expressed himself as follows:—

" The passage from the physics of the brain to the corresponding facts of consciousness is unthinkable. Granted that a definite thought, and a definite molecular action in the brain occur simultaneously, we do not possess the intellectual organ, nor apparently any rudiment of the organ, which would enable us to pass by a process of reasoning from the one phenomenon to the other. They appear together, but we do not know why. Were our minds and senses so expanded, strengthened, and illuminated as to enable us to see and feel the very molecules of the brain; were we capable of following all their motions, all their groupings, all their electric discharges, if such there be, and were we intimately acquainted with the corresponding states of thought and feeling, we should be as far as ever from the solution of the problem, ' How are these physical processes connected with the facts of consciousness?' The chasm between the two classes of phenomena would still remain intellectually impassable."

In his latest work ("An Introduction to the Classification of Animals,") published in 1869, Professor Huxley unhesitatingly adopts the "well founded doctrine, that life is the cause and not the consequence of organization." In his celebrated article " On the Physical Basis of Life," however, he maintains, that life is a property of protoplasm, and that protoplasm owes its properties to the nature and disposition of its molecules. Hence he terms it " the matter of life," and believes that all the physical properties of organized beings are due to the physical properties of protoplasm. So far we might, perhaps, follow him, but he does not stop here. He proceeds to bridge over that chasm which Professor Tyndall has declared to be " intellectually impassable," and, by means which he states to be logical, arrives at the conclusion, that our " *thoughts are the expression of molecular changes in that matter of life which is the source of our other vital phenomena.*" Not having been able to find any clue in Professor Huxley's writings, to the steps by which he passes from those vital phenomena, which consist only, in their last analysis, of movements of particles of matter, to those other phenomena which we term thought, sensation, or consciousness ; but, knowing that so positive an expression of opinion from him will have great weight with many persons, I shall endeavour to show, with as much brevity as is compatible with clearness, that this theory is not only incapable of proof, but is also, as it appears to me, inconsistent with accurate conceptions of molecular physics. To do this, and in order further to develop my views, I shall have to give a brief sketch of the most recent speculations and discoveries, as to the ultimate nature and constitution of matter.

The Nature of Matter.

It has been long seen by the best thinkers on the subject, that atoms,—considered as minute solid bodies from which emanate the attractive and repulsive forces which give what we term matter its properties,—could serve no purpose whatever; since it is universally admitted that the supposed atoms never touch each other, and it cannot be conceived that these homogeneous, indivisible, solid units, are themselves the ultimate *cause* of the forces that emanate from their centres. As, therefore, none of the properties of matter can be due to the atoms themselves, but only to the forces which emanate from the points in space indicated by the atomic centres, it is logical continually to diminish their size till they vanish, leaving only localized centres of force to represent them. Of the various attempts that have been made to show how the properties of matter may be due to such modified atoms (considered as mere centres of force), the most successful, because the simplest and the most logical, is that of Mr. Bayma, who, in his "Molecular Mechanics," has demonstrated how, from the simple assumption of such centres having attractive and repulsive forces (both varying according to the same law of the inverse squares as gravitation), and by grouping them in symmetrical figures, consisting of a repulsive centre, an attractive nucleus, and one or more repulsive envelopes, we may explain all the general properties of matter; and, by more and more complex arrangements, even

the special chemical, electrical, and magnetic properties
of special forms of matter.*   Each chemical element
will thus consist of a molecule formed of simple atoms,
(or as Mr. Bayma terms them to avoid confusion,
" material elements ") in greater or less number and
of more or less complex arrangement; which molecule
is in stable equilibrium, but liable to be changed in
form by the attractive or repulsive influences of differently constituted molecules, constituting the phenomena
of chemical combination, and resulting in new forms
of molecule of greater complexity and more or less
stability.

Those organic compounds of which organized beings
are built up, consist, as is well known, of matter of an
extreme complexity and great instability; whence result the changes of form to which it is continually
subject.   This view enables us to comprehend the *possibility*, of the phenomena of vegetative life being due to

* Mr. Bayma's work, entitled " The Elements of Molecular
Mechanics," was published in 1866, and has received less
attention than it deserves. It is characterised by great
lucidity, by logical arrangement, and by comparatively simple
geometrical and algebraical demonstrations, so that it may
be understood and appreciated with a very moderate knowledge of mathematics. It consists of a series of Propositions,
deduced from the known properties of matter; from these
are derived a number of Theorems, by whose help the more
complicated Problems are solved. Nothing is taken for
granted throughout the work, and the only valid mode of
escaping from its conclusions is, by either disproving the
fundamental Propositions, or by detecting fallacies in the
subsequent reasoning.

an almost infinite complexity of molecular combinations, subject to definite changes under the stimuli of heat, moisture, light, electricity, and probably some unknown forces. But this greater and greater complexity, even if carried to an infinite extent, cannot, of itself, have the slightest tendency to originate consciousness in such molecules or groups of molecules. If a material element, or a combination of a thousand material elements in a molecule, are alike unconscious, it is impossible for us to believe, that the mere addition of one, two, or a thousand other material elements to form a more complex molecule, could in any way tend to produce a self-conscious existence. The things are radically distinct. To say that mind is a product or function of protoplasm, or of its molecular changes, is to use words to which we can attach no clear conception. You cannot have, in the whole, what does not exist in any of the parts ; and those who argue thus should put forth a definite conception of matter, with clearly enunciated properties, and show, that the necessary result of a certain complex arrangement of the elements or atoms of that matter, will be the production of self-consciousness. There is no escape from this dilemma,—either all matter is conscious, or consciousness is something distinct from matter, and in the latter case, its presence in material forms is a proof of the existence of conscious beings, outside of, and independent of, what we term matter. *(Note B.)*

Matter is Force.—The foregoing considerations lead us to the very important conclusion, that matter is essentially force, and nothing but force; that matter, as popularly understood, does not exist, and is, in fact, philosophically inconceivable. When we touch matter, we only really experience sensations of resistance, implying repulsive force; and no other sense can give us such apparently solid proofs of the reality of matter, as touch does. This conclusion, if kept constantly present in the mind, will be found to have a most important bearing on almost every high scientific and philosophical problem, and especially on such as relate to our own conscious existence.

All Force is probably Will-Force.—If we are satisfied that force or forces are all that exist in the material universe, we are next led to enquire what is force? We are acquainted with two radically distinct or apparently distinct kinds of force—the first consists of the primary forces of nature, such as gravitation, cohesion, repulsion, heat, electricity, &c. ; the second is our own will-force. Many persons will at once deny that the latter exists. It will be said, that it is a mere transformation of the primary forces before alluded to ; that the correlation of forces includes those of animal life, and that *will* itself is but the result of molecular change in the brain. I think, however, that it can be shown, that this latter assertion has neither been proved, nor even been proved to be possible; and that in making it, a great leap in the dark has been taken from the known to the unknown. It may be at once admitted that the *muscular force* of animals and men, is merely the transformed energy derived from the primary forces of nature. So much has been, if not rigidly proved, yet rendered highly probable, and it is in perfect accordance with all our knowledge of natural forces and natural laws. But it cannot be contended that the physiological balance-sheet has ever been so accurately struck, that we are entitled to say, not one-thousandth part of a grain more of force has been exerted by any organized body or in any part of it, than has been derived from the known primary forces of the material world. If that were so, it would absolutely negative the existence of will; for if will is anything, it is a power that *directs* the action of the forces stored up in the body, and it is not conceivable that this *direction* can take place, without the exercise of some force in some part of the organism. However delicately a machine may be constructed, with the most exquisitely contrived detents to release a weight or spring by the exertion of the smallest possible amount of force, *some* external force will always be required; so, in the animal machine, however minute may be the changes required in the cells or fibres of the brain, to set in motion the nerve currents which loosen or excite the pent up forces of certain muscles, *some force* must be required to effect those changes. If it is said, "those changes are automatic, and are set in motion by external causes," then one essential part of our consciousness, a certain amount of freedom in willing, is annihilated; and it is inconceivable how or why there should have arisen any consciousness or any apparent will, in such purely automatic organisms. If this were so, our apparent WILL would be a delusion, and Professor Huxley's belief—"that our volition counts for something as a condition of the course of events," would be fallacious, since our volition would then be but one link in the chain of events, counting for neither more nor less than any other link whatever.

If, therefore, we have traced one force, however minute, to an origin in our own WILL, while we have no knowledge of any other primary cause of force, it does not seem an improbable conclusion that all force may be will-force; and thus, that the whole universe, is not merely dependent on, but actually *is*, the WILL of higher intelligences or of one Supreme Intelligence. It has been often said that the true poet is a seer; and in the noble verse of an American poetess, we find expressed, what may prove to be the highest fact of science, the noblest truth of philosophy :

> God of the Granite and the Rose!
>   Soul of the Sparrow and the Bee!
> The mighty tide of Being flows
>   Through countless channels, Lord, from thee.
> It leaps to life in grass and flowers,
>   Through every grade of being runs,
> While from Creation's radiant towers
>   Its glory flames in Stars and Suns.

Conclusion.

These speculations are usually held to be far beyond the bounds of science; but they appear to me to be more legitimate deductions from the facts of science, than those which consist in reducing the whole universe, not merely to matter, but to matter conceived and defined so as to be philosophically inconceivable. It is surely a great step in advance, to get rid of the notion that *matter* is a thing of itself, which can exist *per se*, and must have been eternal, since it is supposed to be indestructible and uncreated,—that force, or the forces of nature, are another thing, given or added to matter, or else its necessary properties,—and that mind is yet another thing, either a product of this matter and its supposed inherent forces, or distinct from and co-existent with it;—and to be able to substitute for this complicated theory, which leads to endless dilemmas and contradictions, the far simpler and more consistent belief, that matter, as an entity distinct from force, does not exist; and that FORCE is a product of MIND. Philosophy had long demonstrated our incapacity to prove the existence of matter, as usually conceived; while it admitted the demonstration to each of us of our own self-conscious, ideal existence. Science has now worked its way up to the same result, and this agreement between them should give us some confidence in their combined teaching.

The view we have now arrived at seems to me more grand and sublime, as well as far simpler, than any other. It exhibits the universe, as a universe of intelligence and will-power; and by enabling us to rid ourselves of the impossibility of thinking of mind, but as connected with our old notions of matter, opens up infinite possibilities of existence, connected with infinitely varied manifestations of force, totally distinct from, yet as real as, what we term matter.

The grand law of continuity which we see pervading our universe, would lead us to infer infinite gradations of existence, and to people all space with intelligence and will-power ; and, if so, we have no difficulty in believing that for so noble a purpose as the progressive development of higher and higher intelligences, those primal and general will-forces, which have sufficed for the production of the lower animals, should have been guided into new channels and made to converge in definite directions. And if, as seems to me probable, this has been done, I cannot admit that it in any degree affects the truth or generality of Mr. Darwin's great discovery. It merely shows, that the laws of organic development have been occasionally used for a special end, just as man uses them for his special ends ; and, I do not see that the law of " natural selection " can be said to be disproved, if it can be shown that man does not owe his entire physical and mental development to its unaided action, any more than it is disproved by the existence of the poodle or the pouter pigeon, the production of which may have been equally beyond its undirected power.

The objections which in this essay I have taken, to the view,—that the same law which appears to have sufficed for the development of animals, has been alone the cause of man's superior physical and mental nature, —will, I have no doubt, be over-ruled and explained away. But I venture to think they will nevertheless maintain their ground, and that they can only be met by the discovery of new facts or new laws, of a nature very different from any yet known to us. I can only hope that my treatment of the subject, though necessarily very meagre, has been clear and intelligible; and that it may prove suggestive, both to the opponents and to the upholders of the theory of Natural Selection.

NOTES.

NOTE A. (*Page* 360.)

Some of my critics seem quite to have misunderstood my meaning in this part of the argument. They have accused me of unnecessarily and unphilosophically appealing to "first causes" in order to get over a difficulty—of believing that "our brains are made by God and our lungs by natural selection;" and that, in point of fact, "man is God's domestic animal." An eminent French critic, M. Claparède, makes me continually call in the aid of—"*une Force supérieure*," the capital F, meaning I imagine that this "higher Force" is the Deity. I can only explain this misconception by the incapacity of the modern cultivated mind to realise the existence of any higher intelligence between itself and Deity. Angels and archangels, spirits and demons, have been so long banished from our belief as to have become actually unthinkable as actual existences, and nothing in modern philosophy takes their place. Yet the grand law of "continuity," the last outcome of modern science, which seems absolute throughout the realms of matter, force, and mind, so far as we can explore them, cannot surely fail to be true beyond the narrow sphere of our vision, and leave an infinite chasm between man and the Great Mind of the universe. Such a supposition seems to me in the highest degree improbable.

Now, in referring to the origin of man, and its possible determining causes, I have used the words "some other power"—"some intelligent power"—"a superior intelligence"—"a controlling intelligence," and only in reference to the origin of universal forces and laws have I spoken of the will or power of "one Supreme Intelligence." These are the only expressions I have used in alluding to the power which I believe has acted in the case of man, and they were purposely chosen to show, that I reject the hypothesis of "first causes" for any and every *special* effect in the universe, except in the same sense that the action of man or of any other intelligent being is a first cause. In using such terms I wished to show plainly, that I contemplated the possibility that the development of the essentially human portions of man's structure and intellect may have been determined by the directing influence of some higher intelligent beings, acting through natural and universal laws. A belief of this nature may or may not have a foundation, but it is an intelligible theory, and is not, *in its nature*, incapable of proof; and it rests on facts and arguments of an exactly similar kind to those, which would enable a sufficiently powerful intellect to deduce, from the existence on the earth of cultivated plants and domestic animals, the presence of some intelligent being of a higher nature than themselves.

NOTE B. (*Page* 365.)

A friend has suggested that I have not here explained myself sufficiently, and objects, that *life* does not exist in matter any more than *consciousness*, and if the one can be produced by the laws of matter, why may not the other? I reply, that there is a radical difference between the two. Organic or vegetative life consists essentially in chemical transformations and molecular motions, occurring under certain conditions and in a certain order. The matter, and the forces which act upon it, are for the most part known; and if there are any forces engaged in the manifestation of vegetative life yet undiscovered (which is a moot question), we can conceive them as analogous to such forces as heat, electricity, or chemical affinity, with which we are already acquainted. We can thus clearly *conceive* of the transition from dead matter to living matter. A complex mass which suffers decomposition or decay is dead, but if this mass has the power of attracting to itself, from the surrounding medium, matter like that of which it is composed, we have the first rudiment of vegetative life. If the mass can do this for a considerable time, and if its absorption of new matter more than replaces that lost by decomposition, and if it is of such a nature as to resist the mechanical or chemical forces to which it is usually exposed, and to retain a tolerably constant form, we term it a living organism. We can *conceive* an organism to be so constituted, and we can further conceive that any fragments, which may be accidentally broken from it, or which may fall away when its bulk has become too great for the cohesion of all its parts, may begin to increase anew and run the same course as the parent mass. This is growth and reproduction in their simplest forms; and from such a simple beginning it is possible to conceive a series of slight modifications of composition, and of internal and external forces, which should ultimately lead to the development of more complex organisms. The LIFE of such an organism may, perhaps, be nothing added to it, but merely the name we give to the result of a balance of internal and external forces in maintaining the permanence of the form and structure of the individual. The simplest conceivable form of such life would be the dewdrop, which owes its existence to the balance between the condensation of aqueous vapour in the atmosphere and the evaporation of its substance. If either is in excess, it soon ceases to maintain an individual existence. I do not maintain that vegetative life *is* wholly due to such a complex balance of forces, but only that it is *conceivable* as such.

With CONSCIOUSNESS the case is very different. Its phenomena are not comparable with those of any kind of *matter* subjected to any of the known or conceivable *forces* of nature; and we cannot *conceive* a gradual transition from absolute unconsciousness to consciousness, from an unsentient organism to a sentient being. The merest rudiment of sensation or self-consciousness is infinitely removed from absolutely non-sentient or unconscious matter. We can conceive of no physical addition to, or modification of, an unconscious mass which should create consciousness; no step in the series of changes organised matter may undergo,

which should bring in sensation where there was no sensation or power of sensation at the preceding step.   It is
because the things are utterly incomparable and incom-
mensurable that we can only conceive of *sensation* coming
to matter from without, while *life* may be conceived as
merely a specific combination and co-ordination of the matter
and the forces that compose the universe, and with which
we are separately acquainted.   We may admit with Professor
Huxley that *protoplasm* is the " matter of life " and the cause
of organisation, but we cannot admit or conceive that *protoplasm* is the primary source of sensation and consciousness,
or that it can ever of itself become *conscious* in the same
way as we may perhaps conceive that it may become *alive*.

INDEX.